Mathematics:
A Simple Tool for Geologists

T0280256

David Waltham

Royal Holloway University of London
Department of Geology
Egham
Surrey
UK

SECOND EDITION

Blackwell
Publishing

© 2000 by Blackwell Science Ltd,
a Blackwell Publishing company

BLACKWELL PUBLISHING
350 Main Street, Malden, MA 02148-5020, USA
9600 Garsington Road, Oxford OX4 2DQ, UK
550 Swanston Street, Carlton, Victoria 3053, Australia

First edition published 1994 by Chapman and Hall
Second edition published 2000 by Blackwell Science Ltd

5   2006

Library of Congress Cataloging-in-Publication Data

Waltham, David.
    Mathematics : a simple tool for geologists/
David Waltham.—2nd ed.
        p.cm.
    Includes index.
    ISBN 0-632-05345-3
    1. Geology—Mathematics.
I. Title.
QE33.2M3 W35   2000
510—dc21

ISBN-13: 978-0-632-05345-2

A catalogue record for this title is available from the British Library.

Set by Graphicraft Ltd, Hong Kong

For further information on
Blackwell Publishing, visit our website:
www.blackwellpublishing.com

Contents

Preface

To the second edition

Earth science has moved on a great deal in the six years since I finished the first edition of this book. The biggest change has been that geologists' use of personal computers has gone from being common to being virtually universal whilst use of email and the World Wide Web has become similarly ubiquitous. These changes have made acquisition of mathematical skills even more important to students and professionals. A recent survey of 38 UK employers rated data handling the top skill they required in new earth science graduates. At the same time, the core skills important in quantitative science have not changed. Students still need to know how to manipulate equations, calculate geometric relationships, to display large data sets, present meaningful statistical information and deal with processes whose rates are constantly changing. In this new edition I have kept most of the mathematics from the first edition but I have also incorporated computer spreadsheets which, I hope, will both improve your understanding of the maths and make it easier for you to start using this mathematics for solving real problems. Apart from the spreadsheets, the other major changes are that there are now more questions for you to tackle and that several parts, particularly in the statistics and integration chapters, have been re-written to make them more useful and easier to follow.

   The first edition seems to have sold reasonably well and I hope that this is because it met a real need and was found to be helpful. I have had quite a lot of feedback from readers. Most of this was complimentary and all of it was helpful so I'm glad to have this opportunity to thank those people who have written, emailed or spoken to me. If anyone else feels the need to pass on comments please feel free. I'd particularly welcome feedback on the spreadsheets since these are new and can be quickly changed if there are problems.

To the first edition

I had little idea, when I was asked about shortcomings in available textbooks, that I would end up writing a book myself, but here it is and I hope that you find it useful. My immediate comment was that there was no textbook available for teaching mathematics to geologists which was pitched at the right level. There are some good texts but, generally, they are aimed at students who have a good grasp of mathematics and wish to apply their knowledge to geological problems. There are also good books which cover mathematics at a level more appropriate for the majority of students who did not follow mathematics to the end of their secondary schooling. Unfortunately, these books cover mathematics as a subject in its own right, rather than as a tool for application in other subject areas. There are no books specifically aimed at teaching basic mathematical techniques to geology students.

This is my attempt at such a book and my approach is to use geological examples to illustrate the meaning of the mathematics rather than to discuss how mathematics is applied to geology in practice. Having said that, I hope that many of the examples used in this book are useful as well as illustrative. I also hope that this book will encourage more geologists to use simple mathematical arguments when they are appropriate. There is a tendency for many otherwise well trained geologists to avoid mathematics, although application of a few simple mathematical methods would make their work both more convincing and more testable.

Many people have been encouraging and supportive during the writing of this book. First of all, Una-Jane Winfield from Chapman & Hall who originally persuaded me to write this and was always enthusiastic. Unfortunately, Una-Jane left Chapman & Hall before the project was complete and I should therefore also thank Dominic Recaldin who took over this book at Chapman & Hall at very short notice. The most important people of all have been the students to whom I have taught this material over the years and who have been responsible for its mutation from a simple abstract mathematics course into a more relevant geology-oriented course. In particular, the 'class of 92/93' who were guinea-pigs for this book and who spotted several shortcomings. Other problems were spotted by Stuart Hardy and Chris Willacy, postgraduate students of mine at Royal Holloway, University of London. However, the biggest contribution to improving my original drafts was undoubtedly from the reviewers, Andy Swan at Kingston University and particularly Peter Gould at Liverpool University.

David Waltham (d.waltham@gl.rhbnc.ac.uk)
November 1999

1 Mathematics as a tool for solving geological problems

1.1 Introduction

This book is not about specialized geological mathematics. Mostly, this book is about simple mathematics, the sort that many people are introduced to at school. However, such mathematics is frequently poorly understood by geology undergraduates and few students are able to use the maths they know for solving realistic problems. The objectives of this book are to improve understanding of simple mathematics through the use of geological examples and to improve ability to apply mathematics to geological problems.

This is not a formal mathematics textbook. My aim is to try to instil an intuitive feel for maths. I believe that this is more helpful than a rigid, formal treatment since formality can often obscure the underlying simplicity of the ideas.

Although this book concentrates upon standard mathematical procedures, it does contain a few more specialized techniques. The majority of the mathematics encountered by typical undergraduate students is therefore covered here. The exception is, perhaps, statistics which forms a large part of geo-mathematics and which is well covered by many excellent textbooks. The statistics chapter in this book should form a good introduction to the material covered in those more specialized texts.

Mathematics is much more akin to a language than a science. It is a method of communication rather than a body of knowledge. Thus, the best way to approach a book like this is as you would a text on, say, French or German. You are learning how to communicate with people who understand the mathematical language. You are not learning a collection of facts. Another similarity to learning a language is that you must never pass on to the next lesson until you have grasped the current one. If you do, you will get hopelessly lost and demoralized since subsequent chapters will simply make no sense.

So that you know you have understood sufficiently to move on, each chapter is sprinkled with examples for you to attempt. Mostly these are very short and simple. A few, however, are more difficult and are designed to make you think carefully about the maths just discussed. If you are unable to do one, you should read over the preceding paragraphs again and make sure that you have understood everything. If that does not help then get assistance. Each chapter concludes with additional simple questions as well as more wide ranging questions which will test your ability to apply what you have learnt to more realistic problems. Outline answers to most questions are given at the end of the book and more complete answers are given for some of the more difficult problems. Look carefully at these complete answers since they also show how your answers should be set out. I assume, throughout this book, that you have a calculator and know how to use it.

One difficulty that many students have with mathematics is the large number of specialized mathematical words. Sometimes these words are completely new to the student whilst other times they are used in a similar, but somewhat more precise, manner to their everyday meaning. It is impossible to avoid use of such words since they are vital in mathematics. Wherever I introduce such a word it is in bold face (e.g. **jargon**).

This first chapter is about basic tools that are needed in succeeding chapters and will introduce you to the most important ideas needed for application of mathematical principles to geological problems.

1.2 Mathematics as an approximation to reality

Geology is frequently regarded as a **qualitative** (i.e. descriptive) science. Geological discussions often revolve around questions about what happened and in what order. For example, was a particular area under the sea when a given sedimentary rock was deposited and does the erosive surface at the top imply that uplift above sea level occurred subsequently? However, the same geological information can be described **quantitatively** (i.e. by numbers). In the preceding example, how deep was the sea and how long was it before uplift occurred? Geology is also concerned with the influence of one process upon another. How does changing water depth affect sediment type? Once again, it is possible to do this quantitatively by producing equations relating, say, grain size to water depth (unlikely to be very accurate but in principle possible).

Fig. 1.1 Sedimentation at a lake floor. Older sediments are slowly buried by younger deposits.

Figure 1.1 illustrates a situation in which a quantitative description can be attempted. The figure shows a lake within which sediment, suspended in the water, rains down and slowly builds up on the lake floor. Obviously, early deposits will be covered by later ones. This results in a relationship between depth below lake bed and time since deposition; the deeper you go the older the sediments get. Now, if the rate at which sediments settle upon the floor is approximately constant, sediments buried 2 metres below the lake bed are twice as old as sediments buried by 1 metre and sediments buried by 3 metres are three times as old and so on. Thus, if you double depth, you double the age, if you triple depth you triple the age and so on. This means that the sediment age is **proportional** to burial depth. This can be expressed, mathematically, by the equation

$$Age = k.Depth \tag{1.1}$$

where k is a **constant**. Constants are values which don't change within a given problem. The period symbol, '.', is often used in place of '×' to indicate multiplication and so Eqn. (1.1) reads '*Age* equals k multiplied by *depth*' which can equally well be written in the forms

$$Age = k \times Depth \tag{1.1}$$

or

$$Age = k\ Depth \tag{1.1}$$

All of these different forms for Eqn. 1.1 simply indicate that the age of the sediment equals its depth multiplied by a constant. This constant tells us how rapidly sediments accumulate. A large value for k implies that age increases very rapidly as depth increases (i.e. sediments accumulated very slowly). A low value implies that the age increases more slowly (i.e. sediments accumulated more rapidly). In a particular lake it might take 1500 years for each metre of sediment to accumulate. In this case $k = 1500$ years/metre. A lake with a lower sedimentation rate of, say, 3000 years/metre, would have a more rapid increase in age with depth of burial.

Question 1.1 If $k = 1500$ years/metre calculate, using Eqn. 1.1, the age of sediments at depths of 1 metre, 2 metres and 5.3 metres. Repeat the calculations for $k = 3000$ years/metre.

As you see, it is possible to produce mathematical expressions relating geological **variables** to each other. A variable is a quantity which, in a particular problem, can change its value, e.g. the variable called *Age* in Eqn. 1.1 changes when the variable called *Depth* is altered. Are such quantitative descriptions worth bothering with and are the results worth having? Well, it depends! Sometimes such an exercise will not tell you anything you didn't already know. On other occasions the ability to manipulate and combine mathematical expressions can lead to new insight into geological processes. Mathematical expressions also have the great merit of consistency, they always give the same answer when you use the same data (unlike some geologists I know). Finally, mathematical expressions are capable of being definitively tested. An expression can be used to predict a result and that result can then be checked. Equation 1.1 could be used to predict the age at a particular depth and that age could be tested using, say, a geochemical dating method. If the age is very wrong then there is something wrong with the geological or mathematical model, e.g. some important factor is missing.

Unfortunately, it is quite possible for mathematics to give the wrong answer. In fact, mathematical results are rarely 100% correct. Hopefully though, a mathematical relationship is at least approximately true. The lake sediment example, Eqn. 1.1, is a good case. The assumption that sedimentation happened at the same rate throughout deposition and the, hidden, assumption that the sediments are not compacted by the weight of overlying deposits are unlikely to be completely true. However, provided the sedimentation rate does not vary too much and provided sediment compaction is not too extreme, Eqn. 1.1 should be approximately correct. This is all that is necessary for a mathematical formulation to be useful. It is worth keeping this fact, that mathematical expressions are usually approximations, at the back of your mind. People often make the assumption that, because a mathematical expression is being used, the answer must be right. This is simply not true, not even in physics (equations in physics are also approximations to reality although the approximation is usually so good that this can be safely overlooked).

A final general point about using mathematics for solving problems. If you look on any page in this book, or at any mathematical paper in a geological journal, you will see that there is far more text than there is mathematics. A frequent failing in students' use of mathematics is to write down lots of equations with no explanation of what they are or what they mean. The result is an obscure piece of work which nobody else, even the students themselves 6 months later, can understand. It is also a recipe for sloppy or illogical mathematics. A good guide is that there should be rather more English in a piece of mathematics than equations. The object is to tie the mathematics in with the 'real world' it is describing. We are dealing with applied rather than pure mathematics and it is vital that the geological relevance of any mathematics you use is made totally clear. Hence, the aim is to describe the geological context rather than details of the mathematics itself. At the very least, you must describe all of the constants and variables that you use. It is also good practice to number all equations since this makes it easier for yourself, or anybody else, to refer to a particular expression in later discussion. Finally, a clear diagram is usually an important part of a good mathematical explanation.

1.3 Using symbols to represent quantities

The lake sediment equation used the word *Depth* to represent the quantity depth. However, any other symbol would do equally well. Equation 1.1

$$Age = k.Depth \tag{1.1}$$

could be written

$$a = kz \tag{1.2}$$

where a is age, k is the sedimentation constant and z is depth. Alternatively, it could be written

$$\alpha = \kappa\zeta \tag{1.3}$$

where α is age, κ is the sedimentation constant and ζ is depth (see Table 1.1 for a list of Greek letters such as those used here). The point is that it really

Table 1.1 Lower case and upper case letters of the Greek alphabet.

Greek characters	Name
α, A	alpha
β, B	beta
γ, Γ	gamma
δ, Δ	delta
ϵ, E	epsilon
ζ, Z	zeta
η, H	eta
θ, Θ	theta
ι, I	iota
κ, K	kappa
λ, Λ	lambda
μ, M	mu
ν, N	nu
ξ, Ξ	xi
o, O	omicron
π, Π	pi
ρ, P	rho
σ, Σ	sigma
τ, T	tau
υ, Y	upsilon
ϕ, Φ	phi
χ, X	chi
ψ, Ψ	psi
ω, Ω	omega

Symbol	Usual meaning
z	Depth
T	Temperature
t	Time
x	Horizontal distance
ρ	Density
ϕ	Porosity or grain size
θ	An angle
P	Pressure
r	Radius
υ	Velocity
σ	Stress

Table 1.2 Commonly used symbols and their usual meanings.

doesn't matter. Equation 1.3 is just as valid and just as simple as Eqn. 1.1. However, the unfamiliarity of Greek letters can make an equation look rather daunting. The same equation could equally well be written using Hebrew characters, Chinese pictograms, Egyptian hieroglyphics or even some completely new set of symbols. The use of Greek letters extensively throughout mathematics is simply a tradition, although it does have the benefit of doubling the number of symbols available.

Also traditional is the use of particular symbols to represent commonly encountered quantities. A good example is z which is nearly always used to represent depth. A few other common examples are given in Table 1.2 which is far from complete but it should give the general idea. These symbols are so commonly used for these particular variables that people frequently forget to define them. Thus if, in a particular book or paper, the author is discussing crustal temperatures and the symbol T appears, you can be fairly sure that this will represent a temperature even if the author forgets to define it as such. However, this is not good practice and all symbols should normally be defined.

A few symbols also have specialized mathematical meanings such as the Greek letter delta (i.e. Δ or δ) which is used to denote a small change in a variable. If temperature in the lower crust increased, due to some thermal event, by a small amount (say 10°C) this temperature change is given the symbol ΔT (or δT). If the original temperature was T°C the increased temperature is then $(T + \Delta T)$°C. Another, well known, example of reserving a symbol for a particular mathematical purpose is the use of π to denote the number 3.14 159 . . . , again this is such common usage that you will virtually never see π defined in a book or a paper. In this case, however, this is acceptable since this convention is universally adopted throughout mathematics and the sciences. Other examples of specialized mathematical meanings for symbols will be covered as we come across them in this book.

1.4 Subscripts and superscripts

Another feature of mathematical expressions, which some people find confusing, is the use of subscripts and superscripts. Subscripts are usually used to qualify the meaning of a symbol. For example, if T is used in an expression to denote temperature as a function of depth in the earth, then T_0 might well be used to denote the temperature at the Earth's surface (i.e. at depth = 0.0 metres). Similarly, if a sandstone and a shale are being compared, their densities might be given the symbols ρ_{sand} and ρ_{shale}, respectively, and their porosities would be given the symbols ϕ_{sand} and ϕ_{shale}. The use of subscripts is no different from the use of any other strange character to denote a quantity. Subscripts simply clarify the meaning of a particular symbol.

In sharp contrast, superscripts (often called **exponents**) have a definite mathematical meaning. A superscript is an instruction to raise a number to a power. Thus a^2 means 'square a', a^3 means 'find the cube of a' and a^n means 'multiply a by itself n times'. Whilst on the subject of raising numbers to a power, it is worth briefly reviewing a couple of simple points that will be needed later on. There are three manipulations in particular that you should be familiar with,

$$x^a x^b = x^{a+b} \tag{1.4}$$

$$x^a/x^b = x^{a-b} \tag{1.5}$$

and

$$(x^a)^b = x^{ab} \tag{1.6}$$

For example, one hundred ($= 10^2$) times one hundred equals ten thousand ($= 10^4$), i.e.

$$10^2 \times 10^2 = 10^{2+2} = 10^4$$

and the cube of four ($4 = 2^2$) is sixty-four ($64 = 2^6$), i.e.

$$(2^2)^3 = 2^{2 \times 3} = 2^6$$

Question 1.2 Simplify and, where possible, evaluate the following expressions

(i) $5^{100}.5^4$;

(ii) $(5^{100})^4$;

(iii) $x^2.x^3$;

(iv) $Depth^2.Depth^3$;

(v) $(T_0^3)^4$ where $T_0 = 10$.

Number	Power of 10
1 000	10^3
10 000	10^4
100 000	10^5
1 000 000	10^6
1 billion	10^9

Table 1.3 Some large numbers expressed as positive powers of ten.

1.5 Very large numbers and very small numbers

Many quantities in geology are very large (e.g. the mass of the Earth) or very small (e.g. the mass of gold in one litre of sea water). It is therefore vital that you understand how to deal with very small or very large numbers. Two ways of talking about such extreme numbers are:

(i) The use of **scientific notation;**

(ii) The use of special, very large or very small, units.

Methods are also required for specifying very small fractions such as the fraction of a rare element contained in a mineral specimen.

Scientific notation (or standard notation) is the most flexible of the methods for discussing the very large and the very small. Table 1.3 shows how various large numbers can be represented by powers of ten. Note that in this book, and in most scientific literature, the American definition for one billion (one thousand million) is used rather than the British (one million million). The quick way to find out which power of ten to use for a particular number is simply to count the number of zeros. One million is 1 followed by 6 zeros and therefore equals 10^6.

This is fine for giving large numbers which are an exact power of ten. What about numbers such as two million? This is easy, two million is two times one million. This number is therefore written

$$2\ 000\ 000 = 2 \times 1\ 000\ 000$$
$$= 2 \times 10^6$$

This is an example of scientific notation for a large number. Other more complex numbers can also be dealt with such as 2 200 000. This is simply 2.2 times one million and can therefore be written as 2.2×10^6. The same number could equally well be thought of as 22 times one hundred thousand leading to $2\ 200\ 000 = 22 \times 10^5$. However, for scientific notation it is usual to have the multiplier falling between one and ten and so the former expression (i.e. 2.2×10^6) is preferred.

Small numbers can be dealt with in a similar manner. Table 1.4 shows how various small numbers are expressed as powers of ten. Again there is a quick

Question 1.3 Express the following numbers in scientific notation:
(i) 1000;
(ii) 2000;
(iii) 2500;
(iv) 2523;
(v) 23 000 000;
(vi) Seven billion.

Table 1.4 Some small numbers expressed as negative powers of ten.

Number	Power of 10
0.001	10^{-3}
0.0 001	10^{-4}
0.00 001	10^{-5}
0.000 001	10^{-6}
1 billionth	10^{-9}

way of determining the power of ten to use. Count the zeros including the zero before the decimal point. For example, 0.0 001 has four zeros in total and is written as 10^{-4}. Don't worry if it is not clear to you why a negative power of ten can be used to express these small numbers, this will be explained further in Chapter 2. For now it is only necessary that you accept that it works. Extending this system to numbers which are not exactly a power of ten is then achieved by introducing a multiplier. For example, 0.0 002 is twice 0.0 001 giving

$$0.0\ 002 = 2 \times 0.0\ 001$$
$$= 2 \times 10^{-4}$$

Another example is 0.0 000 054 which is written

$$0.0\ 000\ 054 = 5.4 \times 0.000\ 001$$
$$= 5.4 \times 10^{-6}$$

Question 1.4   Express the following numbers in scientific notation:
(i) 0.001;
(ii) 0.002;
(iii) 0.0 025;
(iv) 0.002 523;
(v) 0.0 000 023;
(vi) Seven billionths.

Table 1.5 List of prefixes used in the SI system.

Multiple	Prefix	Symbol	Example
10^{-18}	Atto	a	Attometre (am)
10^{-15}	Femto	f	Femtometre (fm)
10^{-12}	Pico	p	Picometre (pm)
10^{-9}	Nano	n	Nanometre (nm)
10^{-6}	Micro	μ	Micrometre (μm)
10^{-3}	Milli	m	Millimetre (mm)
1	No prefix		Metre (m)
10^3	Kilo	k	Kilometre (km)
10^6	Mega	M	Megametre (Mm)
10^9	Giga	G	Gigametre (Gm)
10^{12}	Tera	T	Terametre (Tm)

SI (Système International) units are an alternative to the use of scientific notation. In this system a new unit is introduced for each thousand-fold increase or decrease in size. For example, the basic unit of distance is the metre (m). The next unit up from this is the kilometre (km) which is one thousand times bigger. One thousand kilometres is one million metres and this unit is denoted the megametre (Mm). Continuing on up the sequence we get one billion metres (the gigametre, denoted Gm) and one million million metres (the terametre, denoted Tm). Moving in the opposite direction, one thousandth of a metre is a millimetre (mm) and one millionth of a metre is a micrometre (μm). Finally, we get one thousand millionth of a metre which is called a nanometre (nm). Exactly the same set of prefixes is used for any other SI unit. Thus, the mass units, starting from the very small and increasing one thousand-fold for each step, are the nanogram (ng), the microgram (μg), the milligram (mg), the gram (g), the kilogram (kg), the megagram (Mg), the gigagram (Gg) and the teragram (Tg). These prefixes, and a few extra ones, are summarized in Table 1.5.

It is worth noting that a frequently encountered error in the use of this system is to use 'K' rather than 'k' in, for example, kilometre (i.e. this is written km not Km). A capital K is reserved for the **Kelvin** scale of temperature and thus 'Km' is an abbreviation of 'Kelvin metres' not 'kilometres'. Another point to note is that one very commonly encountered unit, the centimetre (cm), is not an SI unit and its use should normally be avoided.

Apart from common usages such as kilometre and kilogram, the SI method for discussing the very large or the very small is not widely employed in geology. One exception is the use of ky and My for thousands of years and millions of years, respectively (N.B. ka and Ma are also frequently used to denote thousands of years and millions of years). Strictly speaking, these are not SI units since the SI unit of time is the second. Nevertheless, ky, My, ka and

Ma are very commonly used in the earth sciences and therefore need to be understood.

> **Question 1.5** How long, in years, is 31.6 gigaseconds? (Hint: First work out how many seconds there are in a year of 365.24 days.) Using scientific notation, how many seconds is this?

When it comes to denoting very small fractions, the usual approach is a simple extension of the percentage system. In percentage notation, the figure '23%' means 23 parts in every hundred. Thus, if a rock specimen is 23% iron by weight, it contains 23 grams of iron in every 100 grams of rock. However, this is not a convenient system for discussing trace elements present in very small fractions of a per cent. Small proportions can be represented by talking about parts per million (ppm) or parts per billion (ppb). The rock specimen might contain 17 ppb of the element lanthanum. This means that every billion grams will contain 17 grams of lanthanum. It might also contain, say, 10 ppm of gold. Thus every million grams of rock will contain 10 grams of gold (i.e. 10 grams of gold in every metric tonne of rock).

> **Question 1.6** Express 0.01% in ppm.

1.6 Manipulation of numbers in scientific notation

Scientific notation is used frequently both in this book and throughout geological literature. You therefore have to know how to add, subtract, multiply and divide numbers expressed in this way.

The trick with addition or subtraction is to use the same power of ten for all numbers. For example, the net rate of increase in mountain height is given by the rate of uplift, which increases mountain height, minus the rate of erosion which tends to reduce mountain height. If the rate of uplift is 3×10^{-3} m/y whilst the rate of erosion is 5×10^{-4} m/y, the net rate of increase in the mountain height is

$$\text{Rate of increase in height} = \text{Rate of uplift} - \text{Rate of erosion}$$
$$= (3 \times 10^{-3}) - (5 \times 10^{-4}) \tag{1.7}$$

The problem here is that the first number has an exponent of -3 whilst the second has an exponent of -4. However, the rate of erosion can be expressed with an exponent of -3 as follows

$$5 \times 10^{-4} = 0.5 \times 10^{-3}$$

Note that the 5 has been reduced by a factor of ten (to give 0.5) whilst the 10^{-4} has been increased by a factor of ten (to give 10^{-3}). Thus, the overall effect is to leave the value unchanged. Replacing the rate of erosion by this new expression gives

Rate of increase in height $= (3 \times 10^{-3}) - (0.5 \times 10^{-3})$

Once the numbers have been expressed using the same power of ten, the subtraction can be performed

Rate of increase in height $= 2.5 \times 10^{-3}$ m/y.

Question 1.7 Evaluate the following:
(i) $(2.5 \times 10^{109}) + (1.5 \times 10^{109})$;
(ii) $(2.5 \times 10^{109}) + (1.5 \times 10^{108})$;
(iii) $(2.5 \times 10^{211}) - (1.5 \times 10^{211})$;
(iv) $(2.5 \times 10^{211}) - (1.5 \times 10^{210})$.

Multiplication and division are more straightforward. The trick here is to rearrange the expressions so that all the multipliers are together and all the powers of ten are together. Some examples should illustrate this:

$(2.5 \times 10^4) \times (3.0 \times 10^3) = (2.5 \times 3.0) \times (10^4 \times 10^3)$

where, at this point, no calculation has been performed. The multipliers and powers of ten have simply been collected together. The calculations implied by the two bracketed terms can then be evaluated to give

$(2.5 \times 3.0) \times (10^4 \times 10^3) = 7.5 \times 10^7$

Similarly, for division

$(5 \times 10^4)/(2.5 \times 10^3) = (5/2.5) \times (10^4/10^3)$
$$= 2 \times 10^1 = 20$$

Combined examples are also done this way, e.g.

$$\frac{(2.5 \times 10^4) \times (3.0 \times 10^3)}{7.5 \times 10^6} = \frac{2.5 \times 3.0}{7.5} \times \frac{10^4 \times 10^3}{10^6}$$
$$= (7.5/7.5) \times (10^7/10^6)$$
$$= 1.0 \times 10^1$$
$$= 10.0$$

Question 1.8 Evaluate the following:
(i) $(2 \times 10^{200}) \times (3 \times 10^{100})$;
(ii) $(4 \times 10^{110})^2$;
(iii) $(4 \times 10^{107})/(2 \times 10^{107})$;
(iv) $(6 \times 10^{100})/(3 \times 10^{50})$.

Question 1.9 If the mass of the Earth is 5.95×10^{24} kg and the volume is 1.08×10^{21} m³, calculate the average density. (Note that density is mass divided by volume).

I will finish this section with a few words of warning about using calculators for performing these sorts of calculations. There are two problems which this frequently causes. Firstly, many students will write down the results in a similar way to the manner in which they appear on the calculator display. Thus, if the correct answer is 3.01×10^8, this appears in the calculator something like $\boxed{3.01\ \ ^8}$. There is a strong temptation to write this down as 3.01^8 which means 3.01 to the power 8 rather than 3.01 times 10 to the power 8. The second problem is that many students would enter this number by typing the following buttons: $\boxed{3},\boxed{.},\boxed{0},\boxed{1},\boxed{\times},\boxed{1},\boxed{0},\boxed{\text{exp}},\boxed{8}$ which gives a display reading $\boxed{3.01\ \ ^9}$. This is because the correct sequence of buttons should have been: $\boxed{3},\boxed{.},\boxed{0},\boxed{1},\boxed{\text{exp}},\boxed{8}$ since the $\boxed{\text{exp}}$ button means multiply by 10 to the power of the following number. Try entering 3.01×10^8 in the two ways suggested above, and you should see what I mean. In general, I would strongly recommend that you perform calculations involving powers of 10 by using the methods shown in the earlier examples.

1.7 Use consistent units

Whenever a calculation is performed, all values used must be expressed using the same units. For example, in the calculation given above for finding the rate of rise of a mountain

Rate of increase in height = Rate of uplift − Rate of erosion   (1.8)

the rate of uplift and the rate of erosion must be given using the same units. Thus, if the rate of uplift was given as

Rate of uplift = 3×10^{-3} m/y

whilst the rate of erosion was given as

Rate of erosion = 1 m/ky

the calculation cannot be performed using these figures since the first figure has units of metres per year whilst the second has units of metres per thousand years. One of the two figures must be converted to the form of the other. In this case it is probably easiest to rewrite the rate of uplift as

Rate of uplift = 3 m/ky

which is the same as 3×10^{-3} m/y since the amount of uplift in one thousand years is 1000 times more than that in one year (i.e. $10^3 \times 3 \times 10^{-3} = 3$).

Equation 1.6 can then be evaluated to give a net rate of mountain rise equal to 2 m/ky.

Similarly, in the lake sedimentation problem (Eqn. 1.1) all figures must have consistent units. Thus, if the age is quoted in years and the depth is quoted in metres, the sedimentation constant will have units of years/metre. If a depth is given in centimetres whilst k is given in years/metre, the depth must first be converted to metres before the calculation of age is done.

> **Question 1.10** Using Eqn. 1.1 and a sedimentation constant of 1000 years/metre, find the age of sediment buried at a depth of 30 cm.

1.8 Spreadsheets

Mathematics and computers are complimentary tools in quantitative science. Mathematics tells us what to calculate whilst computers are increasingly used to perform the final, usually numerical, calculations. Specialized software is often used to perform computations specific to a particular set of problems. However, some computer packages are much more general and can be used to solve many different problems. Of these more general purpose programs, **spreadsheets** must be the most widespread, useful and easy to use.

Spreadsheets consist of grids into which text, numbers or formulas can be typed. Figure 1.2 shows a very simple spreadsheet which lists the number of sites visited on four successive days of fieldwork. All **cells** contain precisely what you see in Fig. 1.2 (i.e. text or a number) apart from cell B7 which contains the formula = B2 + B3 + B4 + B5, i.e. an instruction to add together the contents of cells B2 to B5. Note that the same result could have been achieved if the formula = *sum* (B2 : B5) had been used instead.

The spreadsheet used to create Fig. 1.2 (it's called *Example.xls*) can be obtained using a **Web Browser** (e.g. Netscape or Internet Explorer) from www.gl.rhbnc.ac.uk (use the *links* button at the top of the page). Alternatively, you can access the same files using anonymous ftp for ftp.gl.rhbnc.ac.uk to log into the *pub* directory. This web site also contains many other spreadsheets (a complete list appears in *Index.xls*). The sheets are associated with various parts of this book and are designed to improve your understanding or to help you apply the mathematics you have learned to your own problems. These spreadsheets have been written using Excel version 7.0 but other packages (e.g. Lotus 123) should be able to read them provided they are reasonably recent versions. Most of these spreadsheets are password protected to prevent you from accidentally corrupting them but, if you are a confident spreadsheet user, you are welcome to alter them in any way you like. All sheets and workbooks are protected using the password *maths for*

	A	B
1	**Day**	**Number**
2	Monday	1
3	Tuesday	3
4	Wednesday	5
5	Thursday	2
6		
7	Total	11

Fig. 1.2 Example spreadsheet. All cells here contain either text or a number apart from cell B7 which contains a formula whose result is 11.

geologists. It is quite possible to use this book without using the spreadsheets but, if you do obtain them, I believe that you will find them helpful.

Many of you probably already know how to use spreadsheets but, even if you do, have a look at *Intro.xls* which will show you several spreadsheet features we will be using later. It will also help you revise what you have learned in Sections 1.2 and 1.6 above.

1.9 Further questions

1.11 If $\Omega_d = 3.1 \times 10^4$ and $\mu = 2.7 \times 10^{-2}$ evaluate $\Delta J = \mu / \Omega_d$

1.12 The Earth gains mass every day due to collision with (mostly very small) meteors. Estimate the increase in the Earth's mass since formation assuming that the rate of collision has been constant and that

$\Delta M = 6 \times 10^5$ kg/day
$A_e = 4.5 \times 10^9$ years

where ΔM is the present-day rate of mass gain and A_e is the age of the Earth. What is this mass gain as a fraction of the total present mass of the Earth, M_e, where

$M_e = 5.95 \times 10^{24}$ kg

Re-express this answer in ppb. Given that the Earth is believed to have formed by a process of accretion, has the rate been constant throughout the Earth's history?

1.13 Calculate the volume of the Earth using the expression

$$V = \frac{4\pi r^3}{3}$$

where r is the Earth's radius (equal to 6.37×10^6 m). Note that this method assumes that the Earth is a perfect sphere.

1.14 How long would it take to travel 100 km at 20 km per hour? The following problem is identical in form:

The North Atlantic Ocean is getting wider at an average rate, v_s, of around 4×10^{-2} m/yr and has a width, w, of approximately 5×10^6 m.

(i) Write an expression giving the age, A, of the North Atlantic in terms of v_s and w assuming the present-day spreading rate is typical of the ocean's entire history.

(ii) Evaluate your expression by substituting the values given above.

1.15 In simple models of mountain formation, the mountain is supported by thickened crust such that

$$\Delta z = h\, \rho_c / \Delta\rho$$

where Δz is the amount of crustal thickness, h is the mountain height, ρ_c is the density of the crust and $\Delta\rho$ is the density contrast between the crust and the underlying mantle. Calculate the increase in crustal thickness under a mountain of height 4×10^3 m if the crustal density is 2.5×10^3 kg/m³ and the density contrast is 500 kg/m³.

2 Common relationships between geological variables

2.1 Introduction

This chapter is about relationships between variables. In the last chapter, the depth and age of sediments in a lake were related by the simple expression

$$\text{Age} = k \times \text{Depth} \tag{1.1}$$

Many other geological variables are also related to each other. For example, the internal temperature of the Earth is related to depth (it gets hotter as you get deeper) and the strength of a rock is related to the pressure applied to it (rocks usually become stronger when compressed). However, the precise nature of the relationship varies from one example to another. For simple cases, expressions similar to Eqn. 1.1 will do. In other cases the relationships are more complex.

In Eqn. 1.1 above, age is a **function** of depth. This implies that any given depth produces a unique value for the age. Any type of relationship in which the value of one variable produces a single, unique, value for another is called a function and this term will be met with repeatedly throughout the remainder of this book.

This chapter is probably the most important in the entire book. It is only the fact that we can use mathematical expressions to relate different geological quantities that makes mathematics useful in geology. In practice, the true relationships between quantities such as depth and temperature are usually so complex that they must be approximated by much simpler ones. This chapter will introduce you to some of the most common of these simple relationships starting with the most simple and common of all, the straight line function.

2.2 The straight line

The straight line equation is possibly the most important mathematical expression found in geology since a very large range of geological problems can be approximated using straight line functions.

Returning to the lake sediment problem of Chapter 1, imagine that the lake completely dried out 1 My ago and that there has been no sedimentation in the lake since that time. Under these circumstances all the sediments are

Depth (m)	Age (My)
0	1
20	1.01
40	1.02
60	1.03
80	1.04
100	1.05

Table 2.1 Age of sediments in a brief out take calculated using Eqn. 2.2.

1 My older than we might otherwise think, i.e. the top sediments are 1 My old rather than recent and sediments at a depth of 1 m are, say, 1 000 500 years old rather than 500 years old. An equation to describe the age of the sediments would now be

$$\text{Age} = (k \times \text{Depth}) + \text{Age of top} \tag{2.1}$$

since, in this expression, each age calculated from '$k \times$ Depth' has the 'Age of top' added to it. Thus, if the sedimentation constant k was 500 y/m and the age of the top sediments is 1 My the sediments have an age of

$$\text{Age} = (500 \times \text{Depth}) + 1\ 000\ 000 \tag{2.2}$$

Sediments buried at a depth of 100 m would then have an age of

$$\begin{aligned}\text{Age} &= (500 \times 100) + 1\ 000\ 000 \\ &= 1\ 050\ 000 \\ &= 1.05\ \text{My}\end{aligned}$$

Repeating this calculation for a range of depths from 0 to 100 m gives the values shown in Table 2.1 and plotted in Fig. 2.1.

Question 2.1 Repeat the above calculation for a depth of 50 m.

As you can see, the resultant graph is a straight line. A straight line is completely specified by just two quantities. Firstly, the point where the line crosses the vertical axis tells us how high up the line is. Secondly, the steepness of the line. A different straight line will either cross the vertical axis at another place or it will be less (or more) steep.

The position where the line crosses the vertical axis is called the **intercept** and has a value of 1 My in the particular case of Fig. 2.1. It is essential, when determining this intercept, that the vertical axis crosses the horizontal axis at the point where the depth is zero. If the vertical axis is anywhere else, the age at which the plotted line crosses the axis will be different (Fig. 2.2). Thus, to specify 'how high up' the straight line is, it would be necessary to give both the intercept and the location of the vertical axis. To avoid this the intercept is

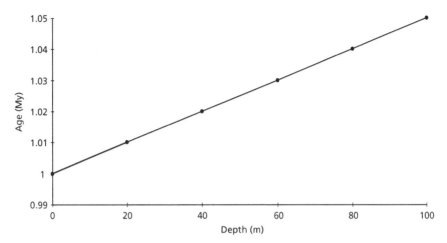

Fig. 2.1 Graph of age versus depth data from Table 2.1.

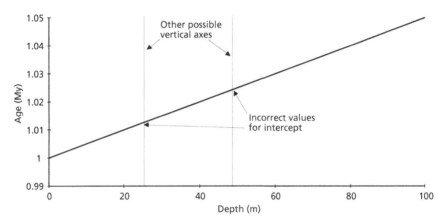

Fig. 2.2 The intercept should be given for a vertical axis passing through zero on the horizontal axis. Any other vertical axis will give a different intercept.

always quoted for a vertical axis which passes through the origin of the horizontal axis.

The second value which characterizes a straight line, the steepness, is called the **gradient** of the line. This is simply a measure of how rapidly the 'height' increases as we move from left to right along the line. The effect of gradient is illustrated by Fig. 2.3. Note that both age and depth alter as we go from point A to point B and that, for a given change in depth, the change in age becomes greater as the steepness of the line increases. Thus, the steepness can be characterized by the increase in age produced by a given increase in depth. For simplicity, we can fix the depth increase as being 1 m, i.e. the gradient is defined as the increase in age produced by drilling into the sediments by an

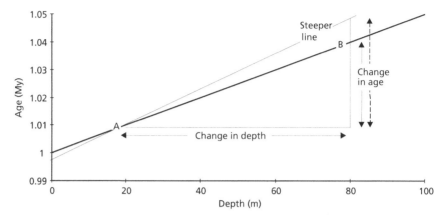

Fig. 2.3 The age and depth alter as we go from point A to point B. However, the age alters more for the steeper line given the same change in depth.

additional 1 m. However, the points A and B in Fig. 2.3 are not necessarily 1 m apart so we must modify the observed change in age between points A and B by dividing it by the distance between them. Thus, the gradient is given by

Gradient = (Change in Age)/(Change in depth)   (2.3)

For example, point A is at a depth of 20 m and an age of 1.01 My whilst point B is at a depth of 80 m and an age of 1.04 My. Thus, the change in depth is 60 m and the change in age is 0.03 My (= 30 000 years) giving a gradient of

Gradient = 30 000/60 = 500 y m⁻¹

There are several points to note about this answer. It does not matter which two points are chosen, the same answer would result if, for example, depths of 0 and 50 m had been chosen for the points A and B. Secondly, the 'units' for the answer of 'years/metre' occur because the top line of the calculation is in years (30 000 years) and the bottom line is in metres (60 m). Hence the calculation involves years divided by metres giving an answer in years/metre. This procedure for finding the units of an answer will be covered in more detail in Chapter 3.

> **Question 2.2** Calculate the gradient of the straight line in Fig. 2.3 using the point A again (depth = 20 m, age = 1.01 My) and the point at a depth of 100 m and age of 1.05 My.

The answer of 500 y m⁻¹ is not only a measure of the steepness of the line. This gradient tells us that each metre of sediment takes 500 years to accumulate (i.e. 500 years per metre). This is, of course, our sedimentation constant.

Table 2.2 Measured ages and depths for sediments in a lake bed.

Depth (m)	Age (years)
0.5	1 020
1.3	2 376
2.47	5 008
4.9	10 203
8.2	15 986

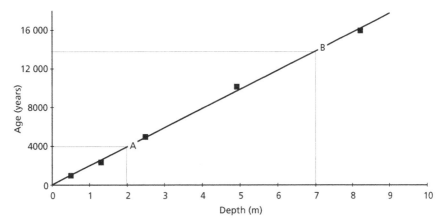

Fig. 2.4 A graph showing the sediment age versus depth data from Table 2.2.

It should now be clear that, for the lake sedimentation example, the intercept on a graph of age against depth tells us the age of the top sediments whilst the gradient tells us the rate of accumulation. In fact, rather than obtaining the straight line graph from given values for intercept and gradient, it is more likely that these quantities will be estimated by fitting a straight line to a graph of some depth/age data. Consider the age versus depth data shown in Table 2.2. These figures might, for example, have been obtained by taking cores from a lake bottom and dating them using the radiocarbon method (this is a geochemical method for estimating the age of organic remains, the details of which are beyond the scope of this book). Figure 2.4 shows a graph of these data together with a 'best-fit' straight line which passes very close to all of the points.

In this example, the intercept value is not significantly different from zero. Thus, in this lake, sedimentation is continuing at the present day and at the same rate as in the past. The gradient of the line can be found by assessing the points A and B shown. The point A lies at a depth of 2 m and an age of 4000 years whilst the point B has a depth of 7 m and an age of 14 000 years. Thus, the change in depth is 5 m and the change in age is 10 000 years giving a gradient of

Gradient = 10 000/5 = 2000 y m^{-1}

i.e. each metre of sediment took 2000 years to accumulate or, equivalently, the sedimentation constant k = 2000 y m^{-1}.

Question 2.3 Given the following depth/age data from a dried-up lake bed, estimate the rate of sedimentation and how long ago the lake dried out.

Depth (m)	Age (years)
6	570 000
10	580 000
18	615 000
20	620 000

It is now time to move from the specific example of lake bottom sedimentation to more general expressions. If we start with our lake sediment equation,

$$\text{Age} = k \times \text{Depth} + \text{Age of top} \tag{2.1}$$

this is not a general equation since the terms in Eqn. 2.1 (i.e. 'Age', '$k \times$ Depth', 'Age of top') refer to specific quantities involved in the sedimentation problem. The simplest way to arrive at a more general expression is to replace each of the terms in the equation by new symbols which do not have specific meanings. Thus we can replace the specific variable 'Age' by the general variable y. Similarly, the 'Depth' can be replaced by the general variable x and the constants k and 'Age of top' can be replaced by the general constants m and c. This procedure results in a general form for the equation of a straight line of

$$y = mx + c \tag{2.4}$$

where y is plotted along the vertical axis and x is plotted along the horizontal axis. Note that using the new symbols y, m, x and c was an entirely arbitrary choice. Any other set of symbols could have been chosen. For example, the equation

$$\alpha = \beta\gamma + \delta$$

is also a straight line equation, provided α and γ are variables and β and δ are constants, since it is of the same form as Eqns. 2.1 and 2.4. However, the particular symbols used in Eqn. 2.4 are traditionally used for the general form of the equation of a straight line and I have stuck to that convention.

A graph of Eqn. 2.4 is a straight line which has an intercept of c and gradient of m (Fig. 2.5). In this figure, Δy (pronounced 'delta y', this means little bit

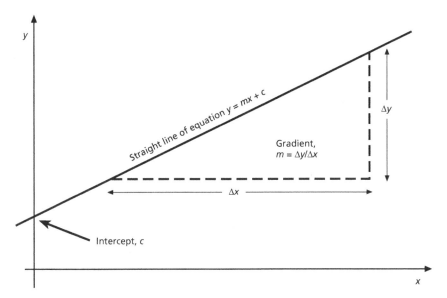

Fig. 2.5 The general form of the equation of a straight line, $y = mx + c$.

of y) is the change in height produced by moving a horizontal distance Δx along the line. Note that, in the illustrated example, y increases for an increase in x since the graph slopes up to the right. Thus Δy and Δx are positive and so is the gradient. On the other hand, if the line sloped downwards to the right, y would decrease as x increases. Thus, Δy would be negative giving a negative gradient. As a general rule, lines that increase in height towards the right have a positive gradient whilst lines that decrease in height have negative gradients.

Spreadsheet *S_line.xls* should help to make this idea clearer. This sheet plots a straight line using Eqn. 2.4. You can alter the value of the gradient, m, and intercept, c, and observe the effect upon the resulting straight line.

The relationship between the general equation for a straight line (Eqn. 2.4) and specific real cases, should be made clearer by one last example. It is well known that temperature increases with depth in the Earth and, for depths of less than around 100 km, it is a good approximation to assume that a plot of temperature against depth should be a straight line. The intercept of such a graph is, by definition, the temperature at zero depth, i.e. the surface temperature. This value will vary considerably from tropical to polar locations but a typical value might be 10°C. The gradient of the line, i.e. the rate at which temperature increases with depth, also varies from one location to another since geologically active areas have very different gradients from old, stable, continental areas. However, values around 20°C km^{-1} are not unusual, i.e. the temperature increases by 20°C for an increase in depth of

1 km. To summarize, temperature plotted against depth gives a straight line characterized by the local temperature gradient and an intercept equal to the local surface temperature. A general expression for how temperature varies with depth at a particular location would therefore be

$$Temperature = (Gradient \times Depth) + Surface\ temperature \qquad (2.5)$$

(cf. $y = mx + c$). For the specific case of an intercept of 10°C and a gradient of 20°C km^{-1} this would yield

$$Temperature = (20 \times Depth) + 10$$

Thus, at a depth of 40 km the temperature is 810°C.

Question 2.4 Rocks usually increase in strength, τ, when compressed. This strength is defined as the shearing (= sideways) pressure necessary for a particular rock specimen to break. The standard units of pressure are pascals. If τ increases by m pascals for each additional pascal of normal pressure (i.e. compressive pressure) and if the strength when not compressed is τ_0, write an equation for how τ varies with normal pressure P. Sketch a graph of this function.

There are many other examples in geology of the use of straight line functions. However, many geological phenomena are not well represented by straight lines and more complex expressions must be used. Some of the more common alternatives are described in the remainder of this chapter.

2.3 Quadratic equations

The linear temperature with depth relationship discussed in the last section breaks down badly for depths much greater than 100 km. For example, at the centre of the Earth the depth is approximately 6360 km so that a surface temperature of 10°C and a gradient of 20°C km^{-1} would predict a temperature of

$$
\begin{aligned}
Temperature &= (Gradient \times Depth) + Surface\ temperature \\
&= (20 \times 6360) + 10 \\
&= 127\ 210°C \qquad (2.5)
\end{aligned}
$$

In fact, the temperature in the Earth's core is only about 4300°C. Table 2.3 lists the approximate temperature in the Earth, at various depths, based upon geophysical and geochemical measurements.

The problem is that the temperature near the surface rises much more rapidly than it does deeper in the Earth, e.g. over 1000 degrees in the first 100 km but only 350 degrees in the following 300 km. Note that the temperature is virtually constant within the inner core (i.e. from 5100 km to the Earth's centre). Any attempt to extrapolate down to the core using the large

Table 2.3 Temperature at various depths in the Earth as determined from geophysical measurements.

Depth (km)	Temperature (°C)
0	10
100	1150
400	1500
700	1900
2800	3700
5100	4300
6360	4300

rate of increase in temperature near the surface is bound to give a ridiculously large value. A much better approximation is

$$Temperature = (-8.255 \times 10^{-5})z^2 + 1.05z + 1110 \tag{2.6}$$

where z is the depth in kilometres. This equation contains three terms (i.e. $(-8.255 \times 10^{-5})z^2$, $1.05z$ and 1110) each of which is calculated separately before adding them together. Thus, at a depth $z = 5100$ km the temperature is

$$Temperature = (-8.255 \times 10^{-5} \times 5100 \times 5100) + (1.05 \times 5100) + 1110$$
$$= -2147 + 5355 + 1110 = 4318°C$$

which compares well with the value given in Table 2.3. However, at the Earth's surface, Eqn. 2.6 predicts

$$Temperature = (-8.255 \times 10^{-5} \times 0 \times 0) + (1.05 \times 0) + 1110$$
$$= 1110°C$$

which is certainly not correct. In fact, Eqn. 2.6 is a reasonable approximation to the Earth's internal temperature only for depths greater than around 100 km.

Figure 2.6 shows how the temperature variation predicted by Eqn. 2.6 compares with the temperatures given in Table 2.3. Although the fit is not exact, it is clear that Eqn. 2.6 can be used to calculate an approximate value for the temperature at any given depth below 100 km. Once again we see that mathematical descriptions of geological behaviour are useful approximations rather than exact 'truths'.

Equation 2.6 is a particular example of a **quadratic equation**. The general form is

$$y = ax^2 + bx + c \tag{2.7}$$

where y is a function of x whilst a, b and c are constants. Figure 2.7 shows a selection of specific examples of Eqn. 2.7. Open spreadsheet *Quadrat.xls* to see, in more detail, how the values of a, b and c affect the shape of a quadratic function. Don't forget to try negative as well as positive values.

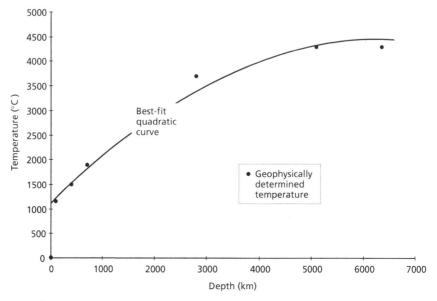

Fig. 2.6 The temperature versus depth data from Table 2.3 compared to a best-fit quadratic function.

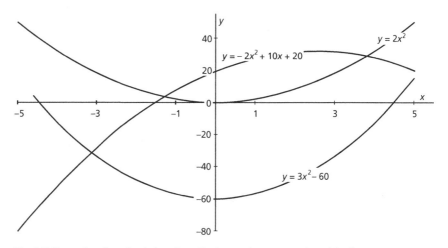

Fig. 2.7 Examples of quadratic functions. Each curve has an equation of the form $y = ax^2 + bx + c$ but the values of a, b and c differ between the curves.

Comparing the general equation (Eqn. 2.7) with the temperature profile function (Eqn. 2.6) we can see that y is equivalent to temperature and x is equivalent to depth. In addition, the constants a, b and c have the values: $a = -8.255 \times 10^{-5}$; $b = 1.05$ and $c = 1110$. The ability to compare a particular equation to a standard form and pick out the appropriate values for the constants will be used again in this book, so make sure that you fully understand what has just been done.

Question 2.5 If $f = 2g^2 - 10g + 6$
where f and g are variables, write down values for the constants equivalent to a, b and c in Eqn. 2.7.

2.4 Polynomial functions

It is possible to improve further on the fit of a mathematical expression to our temperature data by using longer expressions. Figure 2.8 compares the temperature data to the function

$$Temperature = az^4 + bz^3 + cz^2 + dz + e \qquad (2.8)$$

with values for the constants of $a = -1.12 \times 10^{-12}$, $b = 2.85 \times 10^{-8}$, $c = -0.000\,310$, $d = 1.64$ and $e = 930$. However, whilst the fit to the data is now much better, particularly between 2000 and 4000 km, the expression itself is becoming more difficult to evaluate. This trade-off between accuracy and simplicity is a frequent occurrence when applying mathematics to specific problems. Note that even this more complex expression does not model the temperature in the shallowest 100 km.

Question 2.6 Compare the temperature predicted by Eqns. 2.6 and 2.8 at a depth of 2800 km. How do these results compare with the true value in Table 2.3?

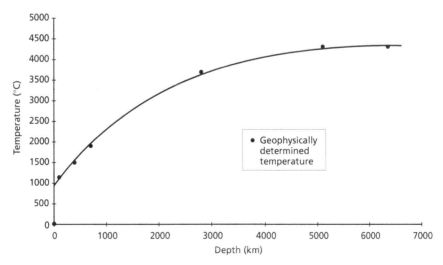

Fig. 2.8 The temperature versus depth data from Table 2.3 compared to a best fit function of the form $Temperature = az^4 + bz^3 + cz^2 + dz + e$ where z is depth.

Expressions like Eqn. 2.8 are known as polynomials (or power series). The general form for these is

$$y = a_0 + a_1x + a_2x^2 + a_3x^3 + \cdots + a_nx^n \qquad (2.9)$$

in which '$\cdots$' indicates a number of terms which have not been explicitly written down. In this expression, a_0, a_1, a_2 etc. are constants and n is an integer giving the power of the last term. For example, if $n = 2$, the expression simplifies to the quadratic function

$$y = a_0 + a_1x + a_2x^2$$

and, if $n = 1$, the expression is a straight line function

$$y = a_0 + a_1x$$

Thus, straight line functions and quadratic functions are special cases of polynomial functions. If $n = 5$, the resulting expression is

$$y = a_0 + a_1x + a_2x^2 + a_3x^3 + a_4x^4 + a_5x^5$$

Spreadsheet *Poly.xls* allows you to investigate how the shape of this function depends upon the **coefficients** a_0, a_1 etc. Remember that you can set some of these coefficients to zero, e.g. set a_2, a_3, a_4 and a_5 all to zero to plot a straight line.

2.5 Negative powers

Look at Table 2.4 which shows 3^n for various values of n. For example, when $n = 1$, $3^n = 3^1 = 3$ and, when $n = 2$, $3^n = 3^2 = 9$. Now, starting at the top ($3^3 = 27$), and moving down the table, each successive result is 1/3 of the result above. For example, 9 is one-third of 27. Continuing this trend down the table, 3^0 should be one-third of 3^1, i.e. $3^0 = 1$. Taking the trend even further, 3^{-1} is 1/3 of 1 (i.e. $3^{-1} = 1/3$). A little further thought should convince you that the more general result is that $3^{-n} = 1/3^n$.

n	3^n
3	27
2	9
1	3
0	1
−1	1/3
−2	1/9
−3	1/27

Table 2.4 The result of raising 3 to the power of the integers between −3 and 3.

The most general result is that x^n divided by x is x^{n-1}. In other words, each decrease of the power by one is achieved by division by x. This is a special case of Eqn. 1.5,

$$x^a/x^b = x^{a-b} \tag{1.5}$$

in which $b = 1$. Thus, Eqn. 1.5 becomes

$$x^a/x = x^{a-1}$$

Now, $x/x = 1.0$ since any number divided by itself is 1.0. In addition, from the discussion above, $x/x = x^{1-1} = x^0$. Thus, $x^0 = 1.0$. This is a very important result: any number raised to the power of zero equals one. (The only exception is $0^0 = 0$.) For example, $2^0 = 1$, $100^0 = 1$, $(-36.4)^0 = 1$ and $\pi^0 = 1$.

This process can be taken a stage further by division of x^0 by x to give x^{-1}. This is the same as dividing 1 by x, i.e. $x^{-1} = 1/x$. Further divisions lead to $x^{-2} = 1/x^2$, $x^{-3} = 1/x^3$, etc. In other words, a number raised to a negative power equals the reciprocal of the same number raised to a positive power. Thus, $(3.5)^{-96} = 1/(3.5)^{96}$, and, in general, $x^{-n} = 1/x^n$.

We are now in a position to see why, in the scientific notation introduced in Chapter 1, numbers smaller than one are expressed using a negative power of 10. Thus, 0.001 is written as 10^{-3} because it equals $1/10^3$ (= 1/1000).

2.6 Fractional powers

The last stage in the generalization of the use of powers is to allow the exponent to be a fraction. Table 2.4 and Fig. 2.9 should help to make this idea more

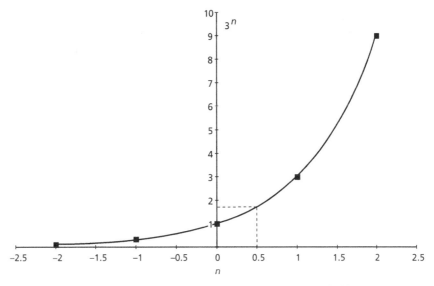

Fig. 2.9 A smooth curve drawn through the data from Table 2.4. The dashed lines indicate the point where $n = 0.5$ from which it can be seen that $3^{1/2} \approx 1.7$.

acceptable. In Fig. 2.9 the points corresponding to $n = -2, -1, 0, 1$ and 2 are plotted as a simple graph which shows that they lie along a smooth curve. This curve can readily be used at points other than $n = -2, -1, 0, 1$ or 2. For example, at $n = 0.5$ the value on the vertical axis is about 1.7. Similarly, at $n = 1.5$ the vertical axis reads approximately 5.2. Thus $3^{0.5}$ is approximately 1.7 and $3^{1.5}$ is about 5.2. A more mathematically formal treatment of this subject is beyond the scope of this book but the main point to learn here is that it is not necessary to use integers when raising a number to a power. Negative fractional exponents are also possible. From Fig. 2.9 it can be seen that $3^{-0.5}$ is around 0.6.

Fractional powers behave in exactly the same way as integer powers. Thus, they obey the equations given in Chapter 1 for manipulating powers, i.e.

$$x^a x^b = x^{a+b} \tag{1.4}$$

$$x^a / x^b = x^{a-b} \tag{1.5}$$

and

$$(x^a)^b = x^{ab} \tag{1.6}$$

For example, from Eqn. 1.4, $x^{0.3} \times x^{0.4} = x^{0.7}$

A direct result of this is that some of these fractional powers have a very simple interpretation. For example, a number raised to the power of 0.5 is the square root of the number ($x^{1/2} = \sqrt{x}$) since $x^{1/2} \times x^{1/2} = x^1$. Similarly, a number raised to the power of one-third results in the cube root ($x^{1/3} = \sqrt[3]{x}$). Figure 2.9 indicated that $3^{1/2}$ is around 1.7; in fact, the square root of 3 is 1.732.

Question 2.7 Draw up a table of 5^n for $n = -2, -1, 0, 1$ and 2. Plot the result. Hence, estimate $1/\sqrt{5}$. In fact, this can be done in two ways. First, estimate it directly from the graph. Secondly, use the graph to estimate $\sqrt{5}$ and then calculate $1/\sqrt{5}$. Compare these answers to each other and to the value given by a calculator.

I will finish this section on polynomial functions and their extensions by using a simple geological example of the use of fractional powers. There are theoretical reasons for expecting water depth, d, in the vicinity of a mid-ocean spreading ridge to depend upon the square root of the distance, x, from the ridge axis according to

$$d = d_0 + ax^{1/2} \tag{2.10}$$

where a is a constant which will depend upon factors such as the spreading rate and d_0 is the depth of the ridge axis. Figure 2.10 shows a comparison between the depths predicted by Eqn. 2.10 and the true water depths in the

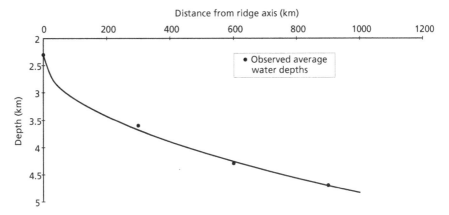

Fig. 2.10 Ocean water depth in the vicinity of the Pacific–Antarctic Ridge. The solid line shows the predicted depth using Eqn. 2.10.

vicinity of the Pacific–Antarctic spreading ridge assuming a value for d_0 of 2.3 km and a value for a of 0.08. Thus, at a distance from the ridge axis of 900 km, Eqn. 2.10 predicts a depth of

$$d = 2.3 + (0.08 \times \sqrt{900}) = 2.3 + (0.08 \times 30) = 2.3 + 2.4 = 4.7 \text{ km}$$

As you can see on Fig. 2.10, the true water depth is indeed very close to this value.

2.7 Allometric growth and exponential functions

Polynomial functions (Section 2.4) and their extensions (Sections 2.5 and 2.6) are extremely versatile and can be used to describe many situations. However, there are situations in which they are not appropriate. The way in which sediments compact as they are buried is a good example.

Water contained in recently deposited sediments is usually squeezed out as the sediments are buried. Thus, sediments start with a relatively large porosity and lose this during burial. A particularly simple approximation for the way in which this happens is to assume that a certain proportion of the water is expelled for a given amount of burial. In a particular case, half of the water might be released when the sediment is buried by 1 km and half of the remaining liquid removed during further burial to 2 km. If the sediment started with a porosity of 0.6 when deposited, the resulting porosity at various depths would be as shown in Table 2.5.

The important point about this example is that porosity always decreases with increasing burial but never actually reaches zero. It would be very difficult to reproduce this using polynomial functions. However, the values in Table 2.5 could be produced by using

Depth (km)	Porosity
0	0.6
1	0.3
2	0.15
3	0.075
4	0.0 375

Table 2.5 Variation in porosity with depth assuming an initial porosity of 0.6 which halves for every kilometre of burial.

$$\phi = 0.6 \times 2^{-z} \qquad\qquad (2.11)$$

in which ϕ is the porosity at a depth z (note that porosity is nearly always denoted by the Greek letter ϕ). For $z = 3$ km, 2^{-z} will be $2^{-3} = 1/8$ and therefore ϕ becomes $0.6 \times 1/8 = 0.075$ as shown in Table 2.5.

Question 2.8 What porosity does Eqn. 2.11 give at a depth of 2 km?

Now, whilst Eqn. 2.11 is similar to those discussed in Section 2.6, the crucial difference is that the variable, z, appears as the exponent in this expression, i.e. the power used varies as z varies. Compare this to Eqn. 2.10 in which the variable, x, is raised to the fixed power 0.5. This subtle difference produces a rather different type of function. Its general form could be expressed as

$$y = ab^{cx} \qquad\qquad (2.12)$$

where y is a function of x whilst a, b and c are constants. Equations such as this are called either **allometric growth laws** or **exponential functions**. This equation does not actually need three separate constants since b^c is itself just another constant ($b^c = d$ say) which means that only two independent constants are needed for the general form of this equation. There are two ways of achieving this. First, simply use $b^c = d$ and write Eqn. 2.12 in the form

$$y = ad^{x} \qquad\qquad (2.13)$$

which has two constants a and d. Alternatively, the constant b in Eqn. 2.12 is fixed to be a particular, convenient, value and c is retained as an independent constant. For example, $b = 10$ may be simple to use in some contexts leading to expressions like

$$y = a \times 10^{cx} \qquad\qquad (2.14)$$

The choice of value to use for b is entirely arbitrary and can be varied to suit the problem. However, 99% of the time a rather peculiar choice for b is made. Normally the value $b \approx 2.718$ is used! This number, denoted by the letter e, is special for reasons which will be touched upon in Chapter 8. For now, it is sufficient to know that it is a very important number in mathematics. It is similar to the number π in that it is **irrational** (i.e. it cannot be expressed exactly as a fraction and, in decimal form, the number goes on for ever); it crops up in many different branches of mathematics and the use of e to denote this number is sufficiently universal that it will not, normally, be defined in most papers or books. Thus, another important form for Eqn. 2.12 is

$$y = a\ e^{cx} \tag{2.15}$$

in which $e \approx 2.718$. An alternative way of writing this is

$$y = a\ \exp(cx) \tag{2.16}$$

which means exactly the same thing as Eqn. 2.15. (N.B. 'exp' here is a single word, it does not mean e times p. In fact, 'exp' is an abbreviation for exponential.)

Spreadsheet *Exp.xls* allows you to plot Eqn. 2.12 for any value of a, b or c. In particular, if c is set to 1 then it models Eqn. 2.13 and, if b is set to 10 then it will plot Eqn. 2.14 for you. To get a standard exponential function (i.e. Eqn. 2.15) you should set b to 2.718 or, alternatively, type the formula $= \exp(1)$ into cell B9.

Equation 2.15 is frequently the form used for modelling the variation in porosity with depth. This leads to expressions such as

$$\phi = \phi_0\ e^{-z/\lambda} \tag{2.17}$$

An example is the best way to illustrate this. If the constants have the values $\phi_0 = 0.7$, $\lambda = 2$ km, the porosity at a depth of 4 km would be

$$
\begin{aligned}
\phi &= 0.7\ \exp(-4/2) \\
&= 0.7\ \exp(-2) \\
&= 0.7 \times 0.135 \\
&= 0.0945
\end{aligned}
$$

Question 2.9 What porosity would this predict for $z = 1$ km?

It is worth spending a little time considering the meaning of the constants ϕ_0 and λ. These do have fairly simple interpretations. First, remember that any number to the power zero equals one. Thus, if the depth is zero, Eqn. 2.17 becomes

$$
\begin{aligned}
\phi &= \phi_0\ \exp(-0/\lambda) \\
&= \phi_0\ \exp(0) \\
&= \phi_0 \times 1.0 \\
&= \phi_0
\end{aligned}
$$

In other words, ϕ_0 is simply the porosity at zero depth. The meaning of λ can be seen by setting z to be λ kilometres. Equation 2.17 then gives

$$\phi = \phi_0 \exp(-\lambda/\lambda)$$
$$= \phi_0 \exp(-1)$$
$$= \phi_0/e$$
$$= \phi_0/2.71$$

i.e. λ is the depth at which the porosity reduces to around one-third of its starting value.

2.8 Logarithms

The logarithmic function is the final type of relationship which will be invest-igated in this chapter. Logarithms solve the problem of how to rearrange equations of the form $y = a^x$ (i.e. exponential or allometric growth functions, Section 2.7) into an equation for x in terms of y. The solution is $x = \log_a(y)$. In other words, logarithms are defined as the **inverse** of exponential func-tions. For example, if $y = 10^3 = 1000$, then $\log_{10}(1000) = 3$. Similarly, $\log_{10}(100\ 000) = 5$ since $100\ 000 = 10^5$. Tables 2.6 and 2.7 make the same point in a slightly different way. Table 2.6 lists the result of raising 10 to the power of various integers, i.e. $10^n = 100$ if $n = 2$ etc. The definition given for logarithms above, implies that Table 2.6 could be rewritten as a table of logarithms simply by swapping around the columns (i.e. Table 2.7).

n	10^n
−2	0.01
−1	0.1
0	1
1	10
2	100
3	1000

Table 2.6 Ten raised the power of the integers between −2 and 3.

Number	Logarithm
0.01	−2
0.1	−1
1	0
10	1
100	2
1000	3

Table 2.7 Table of logarithms produced by swapping the columns in Table 2.6.

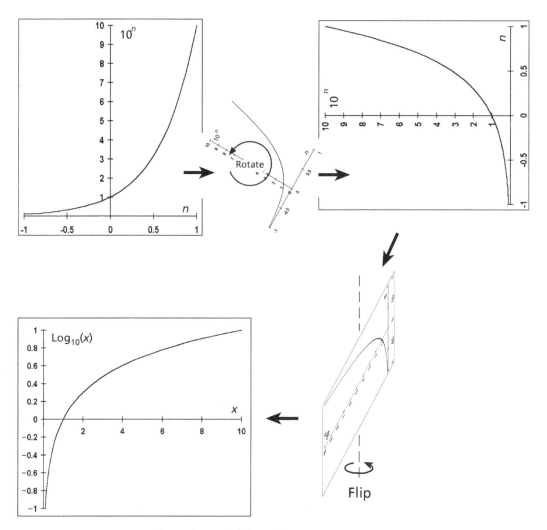

Fig. 2.11 The graph at top left shows the curve $y = 10^n$ using the data from Table 2.6. The graph at bottom left shows the curve $y = \log(x)$ using the data from Table 2.7. This figure shows that the two curves are exactly the same shape since the two functions are very closely related.

Figure 2.11 makes the same point in yet another way. The top left-hand graph shows a graph of 10^n versus n whilst the graph at lower left is of $\log_{10}(x)$ versus x. However, as Fig. 2.11 attempts to make clear, these two functions are very closely related since the two graphs show the same curve just plotted in a different orientation.

Another important point about Fig. 2.11 is that the logarithmic curve never crosses the vertical axis. If this curve was plotted for smaller values of x it would get even steeper and it would never reach the $\log_{10}(x)$ axis. As a consequence, logarithms of negative numbers do not exist. If you try to use a calculator, for example to find $\log_{10}(-2)$, you will get an error message.

Fault length (m)	Number
0.001	10 109
0.01	957
0.1	132
1	11
10	1

Table 2.8 Number of faults of length greater than a given size at a particular outcrop. For example there are 11 faults of length 1 m or longer.

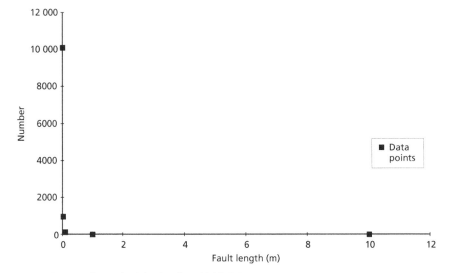

Fig. 2.12 A simple graph of the data from Table 2.8.

There are three main uses for logarithms.

1 Rearranging equations containing exponential functions as discussed above.

2 Reducing exponential and allometric functions to simple straight lines.

3 Compressing large data ranges.

The first two of these are closely related and will be covered in more detail in the next chapter. What about the third use, i.e. compressing large ranges? Fault sizes provide a good example. Faults occur on a vast range of scales from millimetres long to hundreds of kilometres long. Now, in a particular area, the number of faults of different sizes might be something like Table 2.8 in which the fault length is tabulated against the number of faults observed of this size or larger. Note that such tables typically show that small faults are much more common than larger faults. If we attempt to plot these data on a graph, the result is as shown in Fig. 2.12.

This graph is not very helpful because all the points lie on or near the axes. The problem is the large range of values that occurs; some values are very

Table 2.9 Result of taking logarithms of the data in Table 2.8.

Fault length (m)	log(length)	Number	log(number)
0.001	−3	10 109	4.00
0.01	−2	957	2.98
0.1	−1	132	2.12
1	0	11	1.04
10	1	1	0.00

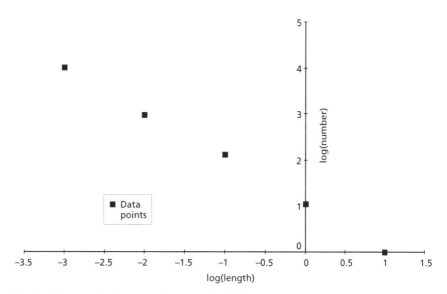

Fig. 2.13 Graph of the logarithm data from Table 2.9.

small whilst some are very large. This makes it impossible to find axes scales which enable all the data to be properly viewed. However, if we add logarithms to the columns in Table 2.8 (to give Table 2.9) and plot these instead, Fig. 2.13 results. Clearly, this graph is much more informative since the data are now spread more evenly across it.

Thus, in summary, a major use for logarithms is for plotting graphs of quantities which vary over large ranges. This is something which occurs very frequently in geological problems.

2.9 Logarithms to other bases

The previous section explained how logarithms were obtained from a table showing the number 10 raised to various powers. However, it is not necessary for the number 10 to be used. If other numbers are used then the result is a logarithm in a different base. An alternative choice might be, for example, a

n	6^n
-2	$1/6^2 = 1/36 = 0.0278$
-1	$1/6^1 = 1/6 = 0.167$
0	$6^0 = 1$
1	$6^1 = 6$
2	$6^2 = 36$

Hence:

x	$\log_6(x)$
0.0278	-2
0.167	-1
1	0
6	1
36	2

Table 2.10 The result of raising 6 to the power of various integers and the resulting table of logarithms obtained by swapping the columns around.

base of 6. Table 2.10 shows the number 6 raised to the power of various integers and the resulting table of logarithms obtained by swapping the columns around.

Note that, in order to indicate that these are logarithms to base 6, a 6 subscript is written after the word 'log'. Logarithms to base 10, i.e. those discussed in section 2.8, are frequently written without this subscript so that, if the subscript is missing, logs to base 10 should be understood. Logarithms to base 10 are sometimes called **common logarithms**.

> **Question 2.10** What number has a logarithm, in base 5, of 2 (i.e. if $\log_5(x) = 2$, what is x)? Hint: Construct a table similar to Table 2.10 but for a base of 5.

A commonly used scale for quantifying sediment grain sizes is called the phi scale and this uses logarithms to base 2. The formal definition of the phi grain size is

$$\phi = -\log_2(d) \tag{2.18}$$

where d is the grain size in millimetres. This is not as complex as it sounds as a table of base 2 logarithms shows (Table 2.11).

To convert phi values into grain sizes in millimetres it is only necessary to start at 1 mm and halve this ϕ times (e.g. for $\phi = 3$, halving 1.0 mm three times gives a grain size of 1/8 mm). A convention that halving a negative number of times means doubling the same number of times must also be used (e.g. for $\phi = -3$, doubling 1.0 mm three times gives a grain size of 8 mm).

Table 2.11 Logarithms to base 2. Note that the log increases by one for each doubling of x and $\log_2(1) = 0$.

x	$\log_2(x)$
0.25	-2
0.5	-1
1	0
2	1
4	2
8	3

However, for grain sizes which are not an integer power of 2 (0.25, 0.5, 1, 2, 4, 8 mm, etc., are all integer powers of 2) this procedure will not work and the formula given by Eqn. 2.18 must be used. Many calculators allow you to do this directly but, if you do not have access to such a calculator, there is a simple recipe for converting between logarithm bases:

$$\log_b(a) = \frac{\log_c(a)}{\log_c(b)} \tag{2.19}$$

Converting a logarithm to the base 2 into a common logarithm should make this clearer. If $b = 2$ and $c = 10$, Eqn. 2.19 becomes

$$\log_2(a) = \frac{\log_{10}(a)}{\log_{10}(2)} = \log_{10}(a)/0.301$$

$$= 3.32 \log_{10}(a) \tag{2.20}$$

i.e. use the common logarithm and then multiply by 3.32. So, a grain size of 2.3 mm would have a phi value given by

$$\phi = -\log_2(2.3)$$
$$= -3.32 \log_{10}(2.3)$$
$$= -3.32 \times 0.362$$
$$= -1.20.$$

To finish this chapter I'll discuss a particularly common base for logarithms, namely logarithms to base e (remember $e \approx 2.718$). These are known as natural logarithms and are denoted either by using the subscript e (natural log of x is written $\log_e(x)$) or, more commonly, it is denoted by ln (i.e. natural log of x is written $\ln(x)$). This type of logarithm probably occurs more frequently than any other, and will be used throughout this book, so it is important to be familiar with its appearance.

Spreadsheet *Log.xls* allows you to plot the common logarithm function, the natural logarithm function and a logarithm function assuming a user-defined base. Using this, for example, you can investigate the $\log_2$ function.

2.10 Further questions

2.11 The following data were taken from the Troll 3.1 well in the Norwegian North Sea.

Depth (cm)	Age (years)
19.75	1 490
407.0	10 510
545.0	11 160
825.0	11 730
1158.0	12 410
1454.0	12 585
2060.0	13 445
2263.0	14 685

By plotting a graph of these data, estimate:
(i) the sedimentation rate for the last 10 000 years;
(ii) the sedimentation rate for the preceding 5000 years;
(iii) the time since sedimentation ceased.
(Data taken from Lehman, S. and Keigwin, L. (1992) Sudden changes in North Atlantic circulation during the last deglaciation. *Nature*, 356, 757–62.)

2.12 As crystals settle out of magmas, element concentrations, C, in the remaining liquid change according to the equation

$$C = C_0 F^{(D-1)}$$

where C_0 is the concentration of the element in the liquid before crystallization began, F is the fraction of liquid remaining and D is a constant (known as the distribution coefficient). Calculate the concentration of an element after 50% crystallization (i.e. $F = 0.5$) if its initial concentration was 200 ppm and $D = 6.5$.

2.13 Radioactive minerals become less active with time according to the equation

$$\ln(a) = \ln(a_0) - \lambda t$$

where a is the radioactivity, a_0 is the initial radioactivity, t is time and λ is a constant which depends upon the mineral. If $a_0 = 1000$ counts per second and $\lambda = 10^{-7}$ y^{-1}, draw up a table and plot a graph of $\ln(a)$ against t for times ranging from 0 to 100 My. From your graph, estimate the age of a specimen which has decayed to $a = 100$ counts per second.

2.14 The variation in gravitational strength with altitude should obey the equation

$$g = g_0 + ah$$

where g is the measured strength of gravity, g_0 is the gravitational strength at sea level, a is a constant and h is height above sea level. However, the presence of metallic ore bodies, volcanic intrusions, etc., tend to increase the local strength of gravity slightly. Thus, real gravitational measurements do not quite obey this expression. Deviations from this equation can therefore be used to indicate the presence of such features. Using the figures given below, plot a graph of g against h and hence estimate g_0 and a. Hence, calculate the deviation of each measurement from its expected value. Plot a graph of this deviation as a function of position and determine the approximate extent of an ore body known to outcrop in this area.

Horizontal position, x (km)	Altitude, h (m)	Gravity, g (m s^{-2})
0.0	150	9.80 945
0.5	100	9.8 097
1.0	170	9.80 949
1.5	200	9.8 094
2.0	150	9.80 955
2.5	130	9.80 951
3.0	120	9.80 954

2.15 The rate of accumulation, p, of carbonate sediments on a reef is given approximately by

$$p = p_0 \exp(-z/Z)$$

where p_0 and Z are constants and z is depth below sea level.
(i) Calculate p at depths of 0, 2, 4, . . . , 20 m if $p_0 = 3$ m ky^{-1} and $Z = 20$ m. Sketch the results.
(ii) Give an interpretation for the constants p_0 and Z.

3 Equations and how to manipulate them

3.1 Introduction

The last chapter introduced many of the more common mathematical functions. It is essential that you know how to manipulate expressions containing combinations of such relationships. Sometimes this will be done in order to simplify the expressions. Sometimes it will be necessary to combine several expressions to produce a new one. Very often the form of an expression is inappropriate for a particular task. Whatever the reason, this is the chapter that tells you how to go about combining, simplifying and rearranging mathematical expressions.

Some of the equations that you will see in this chapter are unfamiliar geophysical or geochemical expressions. However, these will not be derived here because this is not a geophysics or geochemistry text. Enough will be said to allow you to understand the context of the problem.

3.2 Rearranging simple equations

It is very obvious, but it is vitally important to appreciate, that an equation is a mathematical statement in which two expressions equal one another. Look again at the lake bed sediment example from Chapter 1:

$$\text{Age} = k \times \text{Depth} \qquad (1.1)$$

The left-hand expression is very simple, it contains 'age'. The right-hand side is also simple and is the product of k and 'Depth'. The point is that the left- and right-hand sides are stated to be equal and this is what makes 1.1 an equation. The reason that I labour this point is that the golden and unbreakable rule when manipulating equations is that, whatever you do, the left- and right-hand sides must remain equal to one another. This is simply achieved. Whenever you manipulate one side of an equation, you must perform exactly the same operation on the other side. Thus, if you add a constant to one side, you must add the same constant to the other side as well; if you double one side, you must double the other; and so on. For example, given Eqn. 1.1, the following expressions are also true:

$\text{Age} + 3 = (k \times \text{Depth}) + 3$ (i.e. add 3 to both sides);
$2 \times \text{Age} = 2k \times \text{Depth}$ (i.e. double both sides);

√Age = √($k \times$ Depth)   (i.e. square root both sides);
log(Age) = log($k \times$ Depth)   (i.e. take logarithms of both sides).

By combining suitable operations on the two sides of an equation, it is possible to rearrange an equation into another form. As an example, suppose that instead of an equation which tells us the age if we know the depth (i.e. Eqn. 1.1), we actually need an equation which tells us the depth we would need to dig, to reach sediments of a specified age. How do we do this? We must manipulate Eqn. 1.1 to give a new equation which has 'Depth =' on the left-hand side rather than 'Age ='. The problem is that 'Depth' only appears in combination with 'k'; it does not stand on its own. We must, somehow, remove 'k'. Now, if '$k \times$ Depth' is divided by 'k' then we are left with 'Depth'. However, if we do this to the right-hand side of Eqn. 1.1 we must also do this to the left-hand side. This gives

Age/k = Depth

which can obviously be rewritten as

Depth = Age/k (3.1)

which is the expression that we wanted.

The above example is very simple and could probably have been done almost automatically by many readers of this book. However, it is important that you read very carefully through the logic of the above example.

> **Question 3.1** Manipulate Eqn. 1.1 to give an expression for k. If, at a depth of 3 m, the age is 3000 years, use your result to determine the sedimentation constant. (Assuming, of course, that Eqn. 1.1 is valid for the lake bed in question.)

As another example, what about manipulating the slightly more complex lake sediment expression from Chapter 2,

Age = ($k \times$ Depth) + Age of top   (2.1)

to give an expression for depth? The problem is that there is another term on the right-hand side to remove. Should we remove k first or 'Age of top'? In fact, it does not really matter. If we try to remove 'k', by dividing by k as before, the result is

Age/k = Depth + (Age of top/k)   (3.2)

Note that both 'Age' and 'Age of top' now appear over k since all terms must be divided by k. An expression for the depth is then found by subtracting the second term on the right-hand side to give, after swapping left and right sides around,

Depth = (Age/k) – (Age of top/k)   (3.3)

Alternatively, we could have begun by attempting to remove 'Age of top' from the right-hand side of Eqn. 2.1. Subtracting 'Age of top' from Eqn. 2.1 gives

Age – Age of top = k × Depth   (3.4)

Dividing this by k then yields

Depth = (Age – Age of top)/k (3.5)

which is the same as Eqn. 3.3 although written in a slightly different form. (If these look totally different to you, do not worry as it will be explained later in this chapter.)

Question 3.2 Starting with Eqn. 2.1, derive an expression for the age of sediments at the surface of a dried-out lake bed. If the sedimentation constant was 5000 y m^{-1} and, at a depth of 10 m, the age was 60 000 years, determine the age of the surface sediments, this time assuming that Eqn. 2.1 is valid.

Yet another example: how do we know the mass of the Earth? The answer is that we know from the strength of gravity at the Earth's surface. The strength of gravity is measured by the acceleration it causes to a falling body; a strong gravitational pull will accelerate a falling apple, say, more than a weak gravitational pull. Physicists tell us that this gravitational acceleration, g, is related to the Earth's mass, M, by the equation

$g = GM/r^2$ (3.6)

where G is a known physical constant and r is the Earth's radius. To use this equation to estimate the Earth's mass it must be rearranged into an expression for M. Now, G and r^2 can be removed from the right-hand side by dividing by G and multiplying by r^2. Thus, the left-hand side must also be divided by G and multiplied by r^2 to give

$gr^2/G = M$

or, after swapping around

$M = gr^2/G$ (3.7)

The values of all the symbols on the right-hand side of Eqn. 3.7 are known. The gravitational acceleration, g, and the gravitational constant G have both been measured very accurately in physicists' laboratories, whilst the radius of the Earth, r, has been known for more than 400 years (in fact, some ancient Greeks had a pretty good idea too!). The values are

$g = 9.81$ m s^{-1}
$r = 6370$ km $= 6.37 \times 10^6$ m
$G = 6.672 \times 10^{-11}$ m^3 kg^{-1} s^{-2}

Do not worry too much about the units of these numbers but it is important that consistent units are used (Chapter 1), so I have converted the Earth's radius into metres since both g and G are given in units which include metres. Substituting these numbers into Eqn. 3.7 gives

$$M = 9.81 \times (6.37 \times 10^6)^2/(6.672 \times 10^{-11})$$
$$= (9.81 \times 6.37^2/6.672) \times 10^{23}$$
$$= 59.7 \times 10^{23}$$
$$= 5.97 \times 10^{24} \text{ kg}$$

which is about six thousand million million million tonnes. Not bad for a small planet!

3.3 Combining and simplifying equations

If we know the mass of the Earth and can also find its volume, the Earth's average density could be calculated. The volume of the Earth can be estimated using the standard formula for the volume, V, of a sphere of radius r. This is

$$V = 4\pi r^3/3 \tag{3.8}$$

Using the radius of the Earth given above and $\pi \approx 3.142$ gives a volume of

$$V = 4 \times 3.142 \times (6.37 \times 10^6)^3/3$$
$$= (4 \times 3.142 \times 6.37^3/3) \times 10^{18}$$
$$= 1083 \times 10^{18}$$
$$= 1.083 \times 10^{21} \text{ m}^3$$

The density (usually denoted by rho, ρ) is related to mass and volume by

$$\rho = M/V \tag{3.9}$$

Thus, using the mass and volume already found, the average density of the Earth is given by

$$\rho = (5.97 \times 10^{24})/(1.083 \times 10^{21})$$
$$= (5.97/1.08) \times 10^3$$
$$= 5.51 \times 10^3$$
$$= 5510 \text{ kg m}^{-3}$$

and this is more than five times the density of water (which is around 1000 kg m^{-3}).

The above derivation was a little tortuous. It is possible to combine the three expressions (Eqns. 3.7–3.9) into a single expression for density. This means less numerical calculation is necessary and fewer errors will be made. Equation 3.9 tells us to divide mass by volume to give density. The mass is given by Eqn. 3.7 whilst the volume is given by Eqn. 3.8. Therefore we can immediately write down

$$\rho = M/V = \text{Eqn. 3.7/Eqn. 3.8}$$
$$= \frac{gr^2/G}{4\pi r^3/3} \tag{3.10}$$

In other words, it is always possible to replace an expression (e.g. M) by another equal expression (in this case gr^2/G). Since the initial and replacement expressions are equal, the right-hand side of the equation does not change its value and the equation remains true. We now have an expression for density which could be evaluated by substituting the known values for g, r and G. However, Eqn. 3.10 looks a bit daunting. Evaluating it is not any easier than separately evaluating M and V as before. Fortunately, it is possible to simplify. First, we multiply both the top of the right-hand side and the bottom of the right-hand side by G. Since we are multiplying the whole expression then by G/G (which equals 1.0), this has no effect upon the left-hand side. The result is

$$\rho = \frac{G(gr^2/G)}{G(4\pi r^3/3)}$$
$$= \frac{gr^2}{4G\pi r^3/3} \tag{3.11}$$

since the two Gs on the top cancel each other. Now we can do a similar trick to remove the division by 3. Multiplying top and bottom by 3 yields

$$\rho = \frac{3gr^2}{4G\pi r^3} \tag{3.12}$$

Finally, by noting that $r^3 = r^2r$ and cancelling, this expression can be further reduced to

$$\rho = \frac{3gr^2}{4G\pi r^2 r}$$
$$= \frac{3g}{4G\pi r} \tag{3.13}$$

This is much simpler to use than Eqn. 3.10. Let's just check that it gives the right result:

$\rho = 3g/(4G\pi r)$
$\quad = 3 \times 9.81/(4 \times 6.672 \times 10^{-11} \times 3.142 \times 6.37 \times 10^6)$
$\quad = [3 \times 9.81/(4 \times 6.672 \times 3.142 \times 6.37)] \times 10^5$
$\quad = 0.0551 \times 10^5$
$\quad = 5510 \text{ kg m}^{-3}$

as before. Note that this result is about twice the density of typical rocks found at the Earth's surface. We can therefore conclude that the deep Earth must be much denser than the near surface for the average to come out so high.

Question 3.3 Prove that if
$w = 3y/(4z)$ and
$x = 2y/(4z)$ then
$w/x = 1.5$

These, then, are the basic tools for equation manipulation:
(i) you can add, multiply, divide, double, halve, subtract or perform any other operation you like, provided that you do exactly the same to both sides of an equation;
(ii) you can always replace an expression by any other expression which is equal to it.

3.4 Manipulating expressions containing brackets

An important mathematical skill is the ability to use brackets effectively. Sometimes an expression can be made a great deal easier to understand, and easier to use, if brackets are added or removed. Brackets will normally be added into an equation by a procedure called **factorization**, whilst the reverse operation which removes brackets is achieved by multiplying out.

We'll start with a simple problem which is analogous to the algebraic problem of multiplying out brackets. Imagine a region which is known to contain 2 tonnes of recoverable gold and 10 tonnes of recoverable silver in every square kilometre. How much gold and silver could be recovered from 2 km²? Obviously it is 4 tonnes of gold and 20 tonnes of silver. In other words, you multiply the reserves in 1 km² by the number of km².

The following problem in bracket multiplication is identical to the problem above

$$2 \times (2x + 10y) = 4x + 20y \qquad (3.14)$$

The left-hand side says that there are two lots of $(2x + 10y)$ and the right-hand side says that this is the same as $4x$ and $20y$. If you think of x as gold and y as silver you should see the equivalence of the two problems. In practice, all you do is multiply each of the terms inside the bracket by the number outside.

This rather easy example actually contains all the mathematics you need to tackle any other cases. For example,

$$2.3(x + 2y + 4z) = 2.3x + 4.6y + 9.2z \tag{3.15}$$

is calculated exactly the same way as before. Simply multiply each of the terms inside the bracket by the number outside. If the number outside is a symbol rather than a number this leads to examples such as

$$a(x + 2y + 4z) = ax + 2ay + 4az \tag{3.16}$$

which is no different from the earlier examples; simply multiply each term in the brackets by a. A final difficulty is if the number outside the bracket is a more complex expression such as $(a + 3)$ giving

$$(a + 3)(x + 2y + 4z) = (a + 3)x + 2(a + 3)y + 4(a + 3)z \tag{3.17}$$

As you can see, this is still done the same way. However, this time we can take things a stage further. Each of the resulting terms in Eqn. 3.17 is a new problem in multiplying out brackets. For example, the first term on the right-hand side is

$$\begin{aligned}(a + 3)x &= x(a + 3) \\ &= ax + 3x \end{aligned} \tag{3.18}$$

The other terms in Eqn. 3.17 can be similarly multiplied out leading to a final answer of

$$(a + 3)x + 2(a + 3)y + 4(a + 3)z = ax + 3x + 2ay + 6y + 4az + 12z \tag{3.19}$$

Question 3.4 Multiply out the brackets in the following examples.
(i) $5(x + 2y)$;
(ii) $5(x + 2.2y)$;
(iii) $5.5(x + 2y)$;
(iv) $5a(x + 2y)$;
(v) $(x - 2y)(x + 2y)$;
(vi) $(x + 2y)^2$

Question 3.5 Earlier in this chapter I stated that the expression

$$\text{Depth} = (\text{Age}/k) - (\text{Age of top}/k) \tag{3.3}$$

was exactly the same as

$$\text{Depth} = (\text{Age} - \text{Age of top})/k \tag{3.5}$$

Verify this by rewriting Eqn. 3.5 in the form

$$\text{Depth} = (1/k)(\text{Age} - \text{Age of top})$$

and multiplying out the bracket.

Factorization is the reverse process to multiplying out of brackets. For example, Eqn. 3.18 above was

$$x(a + 3) = ax + 3x \tag{3.18}$$

Factorization is the process of writing this the other way around:

$$ax + 3x = x(a + 3) \tag{3.20}$$

Its main use is for simplifying the appearance of more complex expressions and relies upon spotting **common factors**. The trick is to spot that both terms on the left-hand side of Eqn. 3.20 contain a factor x, i.e. they are both equal to some quantity multiplied by x (a times x for the first term and three times x for the second). This x can be taken out as a common factor leaving $a + 3$ inside the bracket. Another, more difficult, example might be

$$3.2xy + 6.4xw + z = ? \tag{3.21}$$

The first two terms have a common factor of $3.2x$ which can therefore be written outside a bracket containing $y + 2w$. Thus, the solution is

$$3.2xy + 6.4xw + z = 3.2x(y + 2w) + z \tag{3.22}$$

Note that the third term does not have any factors in common with the first two and is therefore left alone.

Question 3.6 Factorize $6ax + 3ay$

Factorization can be used to derive an equation for the density of a wet, porous sandstone. This rock will be partly made from sand grains with a density of ρ_s and partly made from water with a density of ρ_w. Hence, the average density of the sample will be somewhere between ρ_s and ρ_w. If the porosity, ϕ, is low, the density will be close to that of the sand grains but if the porosity is higher, then the average density will be a little closer to that of water. More mathematically, take a specimen of this sandstone which has a volume V and mass m. This mass will be made up from the mass of water in the volume plus the mass of the grains, i.e.

$$m = m_w + m_s \tag{3.23}$$

where m_w and m_s are the masses of water and sand respectively. However, the mass of water is given by the product of the volume of water and the density of water whilst m_s is similarly given by the grain density times the volume of grains. Thus, Eqn. 3.23 can be written

$$m = V_w\rho_w + V_s\rho_s \tag{3.24}$$

where V_w and V_s are the volumes of water and sand in the sample. The volume of water is equal to the volume of the sample multiplied by the porosity (e.g. porosity equal to 0.5 implies that half the total volume is water, a porosity of 0.25 implies one-quarter of the total volume is water). Thus

$$V_w = \phi V \tag{3.25}$$

The remaining volume must be sand and therefore

$$V_s = V - \phi V \tag{3.26}$$

Substituting these volumes into Eqn. 3.24 gives

$$m = \phi V \rho_w + (V - \phi V)\rho_s \tag{3.27}$$

Multiplying out the bracket then leads to

$$m = \phi V \rho_w + V \rho_s - \phi V \rho_s \tag{3.28}$$

Now V is common to all the terms on the right-hand side and can therefore be taken out as a common factor to give

$$m = V(\phi \rho_w + \rho_s - \phi \rho_s) \tag{3.29}$$

Finally, dividing by V and performing a further factorization gives the required average density

$$\begin{aligned} \rho &= m/V \\ &= \phi \rho_w + \rho_s - \phi \rho_s \\ &= \phi \rho_w + (1 - \phi)\rho_s \end{aligned} \tag{3.30}$$

Question 3.7 Using Eqn. 3.30, plot a graph of how the density varies as porosity changes from zero to one. Assume

$\rho_w = 1000$ kg m^{-3}
$\rho_s = 2500$ kg m^{-3}

3.5 'Rearranging' quadratic equations

Chapter 2 introduced a quadratic expression for calculating temperature for the deeper parts of the Earth. This was in the form

$$\text{Temperature} = (-8.255 \times 10^{-5})z^2 + 1.05z + 1110 \tag{2.6}$$

How can this be rearranged to allow calculation of the depth for a given temperature, e.g. at what depth is the temperature 2000°C? In fact, such a rearrangement is rather difficult. To solve this problem it is first necessary to discuss a technique called finding the **roots** of a quadratic equation.

The roots of a quadratic equation are the values of the variable which make the quadratic expression equal zero. Figure 3.1 should make this idea clearer.

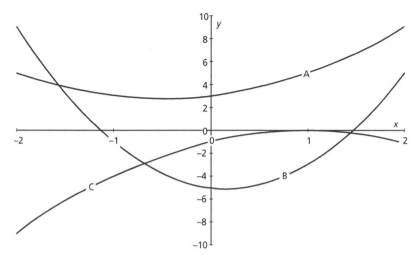

Fig. 3.1 The roots of a quadratic equation are the points where the quadratic curve crosses the horizontal axis. Thus, curve A has no roots, curve B has roots at about $x = -1.1$ and $x = 1.5$, curve C has one root near $x = 1$.

The most general way to find these roots is to use the following method. The roots of the quadratic equation

$$y = ax^2 + bx + c \tag{3.31}$$

are the values of x for which

$$y = 0 \tag{3.32}$$

or, equivalently, the roots are the values of x satisfying

$$0 = ax^2 + bx + c \tag{3.33}$$

From Fig. 3.1 it should be clear that there can be two such values (curve B), or one (curve C) or none (curve A). The values are given by

$$x = \frac{-b \pm \sqrt{b^2 - 4ac}}{2a} \tag{3.34}$$

where $\pm$ means 'either add or subtract'. For example, curve B in Fig. 3.1 had the equation

$$y = 3x^2 - x - 5 \tag{3.35}$$

i.e. $a = 3, b = -1, c = -5$.

Note that both b and c are negative in this example. Substituting these values into Eqn. 3.34 gives

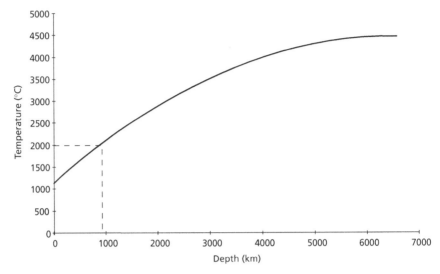

Fig. 3.2 At what depth is the temperature 2000°C?

$$x = [1 \pm \sqrt{(1 + 4 \times 3 \times 5)}]/[2 \times 3]$$
$$= [1 \pm \sqrt{61}]/6$$
$$= [1 + \sqrt{61}]/6 \text{ or } = [1 - \sqrt{61}]/6$$
$$= 1.47 \text{ or } = -1.14 \tag{3.36}$$

which are, indeed, the points where curve B crosses the horizontal axis in Fig. 3.1. If, on the other hand, we look at curve A, this has the equation

$$y = x^2 + x + 3 \tag{3.37}$$

i.e. $a = 1, b = 1, c = 3$.
   Substituting these values into Eqn. 3.34 gives

$$x = [-1 + \sqrt{(1 - 12)}]/2$$
$$= [-1 \pm \sqrt{-11}]/2 \tag{3.38}$$

but $\sqrt{-11}$ has no solution (because you cannot find the square root of a negative number) and therefore there are no roots (which is what Fig. 3.1 shows).

Question 3.8 Find the roots for curve C where
$y = -x^2 + 2x - 1$

We can now attempt the problem of how to find the depth for a particular temperature. Take the case where we wish to know the depth at which the temperature is 2000°C (Fig. 3.2). Substituting this temperature into Eqn. 2.6 gives

$$2000 = (-8.255 \times 10^{-5})z^2 + 1.05z + 1110 \tag{3.39}$$

which is not quite in the form that we need. We have a method for solving equations like 3.33 where the quadratic expression equals zero. Here we have an equation which equals 2000. However, this is simply remedied by subtracting 2000 from both sides to give

$$0 = (-8.255 \times 10^{-5})z^2 + 1.05z - 890 \tag{3.40}$$

which is equivalent to Eqn. 3.33 with values for a, b and c of

$$a = -8.255 \times 10^{-5}, b = 1.05, c = -890.$$

Substituting these into Eqn. 3.34 (and remembering that z is the variable in Eqn. 2.6 not x) gives

$$z = \frac{-1.05 \pm \sqrt{1.05^2 - (4 \times 8.255 \times 10^{-5} \times 890)}}{-2 \times 8.255 \times 10^{-5}} \tag{3.41}$$

which has solutions of $z = 913$ km and $z = 11\ 806$ km. There are two depths because at a depth of 11 806 km, you are about 913 km from the surface on the far side of the Earth.

Question 3.9 Find the depths at which temperature reaches 3000°C.

3.6 Further questions

3.10 The density, ρ, of an air-filled porous rock is given by

$$\rho = \rho_g[1 - (V_p/V)]$$

where ρ_g is the density of the grains making up the rock, V_p is the volume occupied by pore space and V is the total volume. By combining this with Eqn. 3.9 prove that the grain density is given by

$$\rho_g = M/(V - V_p)$$

where M is the mass of the rock sample. Hence, calculate both the average density and the grain density of a sample with a volume of 0.11 m³, a mass of 205 kg and a porosity of 0.32.

3.11 Stokes' law states that the velocity at which a spherical particle suspended in a fluid settles to the bottom is given by

$$v = \frac{2(\rho_p - \rho_f)gr^2}{9\eta}$$

where v is the velocity of descent, ρ_p and ρ_f are the densities of particle and fluid, respectively, g is the acceleration due to gravity, r is the particle radius and η is a property of the fluid known as its **viscosity**. Assuming that grains of different sizes have identical densities, show that the ratio of the settling velocities for two different grain sizes is

$$\frac{v_1}{v_2} = \left(\frac{r_1}{r_2}\right)^2$$

where v_1 and v_2 are the velocities for grains of radius r_1 and r_2, respectively. If a grain of radius 0.1 mm, suspended in a lake, takes 10 days to settle to the lake bottom, how long would it take a grain of radius 1 mm?

3.12 (i) Rearrange

$$0 = ax^2 + bx + c$$

into an equation for b.
(ii) Use your answer from (i) to verify that

$$b^2 - 4ac = a^2x^2 + (c^2/x^2) - 2ac$$

(iii) Verify that the answer to (ii) could be factorized to yield

$$b^2 - 4ac = [ax - (c/x)]^2$$

(Hint: It is easiest to do this by multiplying out the above expression.)
(iv) Use the above answers to verify that one of the roots of a quadratic expression is given by

$$x = \frac{-b + \sqrt{b^2 - 4ac}}{2a}$$

3.13 Check your answers to questions 3.8 and 3.9 by using spreadsheet *Roots.xls*.

4 More advanced equation manipulation

4.1 Introduction

In the last chapter you were introduced to methods for manipulating simple equations. In this chapter we will look at a few, more advanced, techniques for equation manipulation. In particular, I shall discuss manipulation of equations containing exponentials and logarithms. I will also look at the topic of simultaneous equations in which several equations must be manipulated at the same time in order to solve a problem.

This chapter also introduces techniques for checking equations for errors. They may have been wrong in the first place (e.g. due to a printing error) or you may have made mistakes during your manipulations. Either way, it is useful to be able to check that equations are reasonable.

4.2 Expressions involving exponentials and logarithms

In Chapter 2 we looked at expressions involving exponentials such as

$$\phi = \phi_0 \, e^{-z/\lambda} \tag{2.17}$$

for variation in porosity, ϕ, with depth, z. We also looked at logarithms and it was stated that these could be used to recast the graph of an exponential expression into the form of a straight line. This procedure is essential if, for example, you wish to rewrite Eqn. 2.17 to give the depth at which a particular porosity occurs. A little further revision on the properties of logarithms is first needed. The point to remember about logarithms is that they are simply the reverse operation to raising to a power (Section 2.8). From this, it follows that

$$\log_y(y^x) = x \tag{4.1}$$

where y is any base for the logarithm.

For example, $10^2 = 100$ and $\log(100) = 2$, i.e., $\log(10^2) = 2$. Another useful result is that

$$\log(ab) = \log(a) + \log(b) \tag{4.2}$$

where again the log can be to any base. In other words, the logarithm of any two numbers multiplied together is equal to the sum of the logarithms of the numbers. For example,

$$\log(12) = \log(3) + \log(4) \qquad (\text{since } 3 \times 4 = 12)$$

and also

$$\log(12) = \log(6) + \log(2) \qquad (\text{since } 2 \times 6 = 12)$$

or even

$$\log(12) = \log(10) + \log(1.2) \quad (\text{since } 10 \times 1.2 = 12)$$

Equation 4.2 can be generalized into

$$\log(abc \ldots) = \log(a) + \log(b) + \log(c) + \ldots \tag{4.3}$$

i.e. the logarithm of a series of numbers multiplied together equals the sum of their logarithms. For example,

$$\ln(2 \times 3 \times 4.712 \times f) = \ln(2) + \ln(3) + \ln(4.712) + \ln(f)$$

A special case of Eqn. 4.3 is the logarithm of a number raised to a power. In this case we get

$$\log(x^n) = \log(x) + \log(x) + \log(x) + n \text{ terms like this}$$
$$= n \log(x) \tag{4.4}$$

Finally, subtraction of two logarithms is equivalent to division of the **arguments** (the argument of $\log(b)$ is b, the argument of $\ln(f)$ is f and so on). Thus

$$\log(a/b) = \log(a) - \log(b) \tag{4.5}$$

and

$$\ln(a/b) = \ln(a) - \ln(b) \tag{4.6}$$

These rules allow the following transformation of equations such as 2.17.
    Starting with

$$\phi = \phi_0 \, e^{-z/\lambda} \tag{2.17}$$

and taking the natural logarithm of both sides gives

$$\ln(\phi) = \ln(\phi_0 \, e^{-z/\lambda}) \tag{4.7}$$

Now, using the rule about logarithms of products (Eqn. 4.3), leads to

$$\ln(\phi) = \ln(\phi_0) + \ln(e^{-z/\lambda}) \tag{4.8}$$

Finally, Eqn. 4.1 produces

$$\ln(\phi) = \ln(\phi_0) - z/\lambda \tag{4.9}$$

which implies that a graph of $\ln(\phi)$ against z is a straight line of gradient $(-1/\lambda)$ and intercept $\ln(\phi_0)$ (Fig. 4.1).

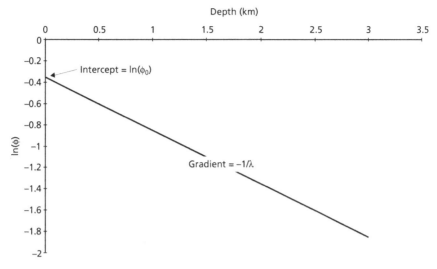

Fig. 4.1 A graph of ln(porosity) against depth assuming Eqn. 2.17, an initial porosity of 0.7 and $\lambda = 2$ km.

We now have an expression which can be rearranged for depth. First add z/λ to both sides of Eqn. 4.9 to give

$$z/\lambda + \ln(\phi) = \ln(\phi_0) \tag{4.10}$$

Then subtract $\ln(\phi)$

$$z/\lambda = \ln(\phi_0) - \ln(\phi) \tag{4.11}$$

Finally, multiply by λ

$$z = \lambda[\ln(\phi_0) - \ln(\phi)] \tag{4.12}$$

Alternatively (using Eqn. 4.6), this can be expressed as

$$z = \lambda \ln(\phi_0/\phi) \tag{4.13}$$

Thus, Eqn. 4.12 or 4.13 can now be used to obtain the depth at which a specific porosity occurs. For example, for $\lambda = 2$ km and $\phi_0 = 0.7$ a porosity of 0.35 would occur at a depth of

$$z = 2 \ln(0.7/0.35)$$
$$= 2 \ln(2)$$
$$= 2 \times 0.693$$
$$= 1.39 \text{ km}$$

A geochemical example may serve to reinforce these ideas. Radioactive dating is based upon the fact that the amount of a radioactive material decreases with time. Thus, if you know how much of the radioactive substance is present now and how much was present when a mineral was formed, it is possible to calculate the time since formation. The amount of material present is given by

$$Q = Q_0 \exp(-t/\lambda) \qquad (4.14)$$

where Q is the amount, Q_0 is the original amount, t is time since formation and λ is a constant which depends upon the radioactive element. This equation is in the same form as the porosity versus depth expression discussed above. It is possible, therefore, to rearrange it in exactly the same way to give

$$t = \lambda \ln(Q_0/Q) \qquad (4.15)$$

which can then be used to calculate the specimen's age. In practice, radioactive dating is usually more complicated than this but the above analysis nevertheless contains the essence of such techniques.

Question 4.1 If $Q_0 = 100$ ppb (parts per billion), $Q = 5$ ppb and $\lambda = 5 \times 10^6$ years, calculate t.

Question 4.2 If the age of a specimen is known to be 100 My and the present concentration of ^{235}U is 10 ppb, find the amount of uranium originally present in the sample. (Use $\lambda = 1.01 \times 10^9$ years for ^{235}U.)

4.3 Simultaneous equations

Perhaps the best way to illustrate **linear simultaneous equations** is to launch straight into a problem. In Chapter 2, I introduced quadratic and polynomial expressions for approximating the temperature versus depth function of the Earth. How were these expressions derived?

To keep this example as simple as possible, I will calculate temperature as a function of distance, r, from the Earth's centre. This is slightly easier than calculating temperature as a function of depth from the surface. The starting point is to observe that the general shape of the temperature versus depth function is symmetric about the Earth's centre, with a steep gradient and low temperature near the Earth's surface (i.e. at large r), and a zero gradient and high temperature at its centre (i.e. at $r = 0$). Now, from the discussion of quadratic equations given in Chapter 2, it should be clear that the expression

$$T = ar^2 + b \qquad (4.16)$$

can be made to have these properties. In this expression, a and b are constants and, in order to have the right properties, b should be the temperature at the Earth's centre and a should be negative (so that temperature reduces as r increases). The problem then becomes simply one of determining values for *a* and *b* which enable Eqn. 4.16 to be a good approximation to the true temperature profile. The next step is to choose two values for *r* where the temperature is known. In this example I will use $T = 4300°C$ at $r = 1260$ km and $T = 1150°C$ at $r = 6260$ km (from Table 2.3 after converting from depth to distance from the Earth's centre). If these are substituted into Eqn. 4.16, the results are

$$4300 = 1\,587\,600\,a + b \tag{4.17}$$

and

$$1150 = 39\,187\,600\,a + b \tag{4.18}$$

This is the system of linear simultaneous equations that we have to solve. The point here is that we have two different equations each of which contains the same unknowns (i.e. *a* and *b*). Now, whenever we have two such equations it is usually possible to combine them to deduce the two unknowns. More generally, whenever we have the same number of unknowns as we have different equations, the equations can usually be combined to calculate the unknown values. In the case of Eqns. 4.17 and 4.18, these may be solved as follows:

1 Rearranging Eqn. 4.17 to give an expression for *b* gives

$$b = 4300 - 1\,587\,600\,a \tag{4.19}$$

2 Substituting this result into Eqn. 4.18 gives

$$1150 = 39\,187\,600\,a + 4300 - 1\,587\,600\,a \tag{4.20}$$

3 Rearranging to give an expression for *a* gives

$$a = (1150 - 4300)/(39\,187\,600 - 1\,587\,600)$$
$$= -3150/37\,600\,000$$
$$= -8.378 \times 10^{-5} \tag{4.21}$$

4 Finally, this result may be substituted back into Eqn. 4.19 to give

$$b = 4300 - (1\,587\,600 \times -8.378 \times 10^{-5})$$
$$= 4300 + 133$$
$$= 4433 \tag{4.22}$$

In some cases, there are short cuts which can be made in this procedure. However, the method given above will always produce a solution if one exists.

Sometimes a solution does not exist. This occurs if the equations are not **linearly independent**. For example, the equations

$$2 = 2a + 3b \tag{4.23}$$

and

$$4 = 4a + 6b \tag{4.24}$$

are not linearly independent since the second equation is simply the first equation after doubling. Another example would be

$$1 = 2a + 3b \tag{4.25}$$

and

$$2 = 2a + 3b \tag{4.26}$$

since these equations cannot both be true simultaneously ($2a + 3b$ cannot equal both 1 and 2). If you try to solve equations which are not linearly independent the result will usually be obvious nonsense. For example, if you rearrange Eqn. 4.25 for a and substitute the result into Eqn. 4.26 then the expression $1 = 2$ results.

Having obtained a solution (e.g. $a = -8.378 \times 10^{-5}$ and $b = 4433$ in Eqns. 4.17 and 4.18), it is a good idea to test the solution by substituting the results back into the original equations. Thus, Eqns. 4.17 and 4.18 become

$$4300 = (1\ 587\ 600 \times -8.378 \times 10^{-5}) + 4433$$
$$= -133 + 4433 = 4300 \tag{4.27}$$

and

$$1150 = (39\ 187\ 600 \times -8.378 \times 10^{-5}) + 4433$$
$$= -3283 + 4433 = 1150 \tag{4.28}$$

The above example was for the case of two equations and two unknowns. In cases where there are more unknowns the procedure is very similar (remember, the number of equations and the number of unknowns must be the same). As an example, take the following case which has three equations and three unknowns:

$$5 = 2x + 4y + 2z \tag{4.29}$$

$$10 = x + y - z \tag{4.30}$$

$$1 = 2x - 2y + 3z \tag{4.31}$$

where x, y and z are the unknowns to be determined. The procedure is:

Step 1 Rearrange any one of the expressions to give one unknown in terms of the other two. It does not matter which equation and which unknown are used. For example, Eqn. 4.30 gives

$$z = x + y - 10 \tag{4.32}$$

Step 2 Substitute into the remaining equations to give a new system of equations. Thus, substituting Eqn. 4.32 into Eqn. 4.29 yields

$$5 = 2x + 4y + 2(x + y - 10)$$
$$= 2x + 4y + 2x + 2y - 20$$
$$= 4x + 6y - 20$$

giving

$$25 = 4x + 6y \qquad (4.33)$$

Similarly, substituting Eqn. 4.32 into Eqn. 4.31 gives

$$1 = 2x - 2y + 3(x + y - 10)$$
$$= 2x - 2y + 3x + 3y - 30$$
$$= 5x + y - 30$$

i.e.

$$31 = 5x + y \qquad (4.34)$$

Step 3 Repeat previous steps for the new, smaller, system of equations. In this case we have a new system consisting of Eqns. 4.33 and 4.34, i.e. two equations and two unknowns. Thus, we may now solve as for the earlier problem. Rearranging Eqn. 4.34 leads to

$$y = 31 - 5x \qquad (4.35)$$

which, upon substitution into Eqn. 4.33, gives

$$25 = 4x + 6(31 - 5x)$$
$$= 4x + 186 - 30x$$
$$= -26x + 186$$

giving

$$-26x = -161$$

i.e.

$$x = 161/26 = 6.1\,923 \qquad (4.36)$$

Step 4 Back substitution. Having obtained one unknown, substitute into earlier equations to determine the others. First, substitute Eqn. 4.36 into Eqn. 4.35 to give

$$y = 31 - (5 \times 6.1\,923)$$
$$= 31 - 30.9\,615 = 0.0\,385 \qquad (4.37)$$

Finally, substitute both Eqns. 4.36 and 4.37 into Eqn. 4.32 to give

$$z = 6.1\,923 + 0.0\,385 - 10$$
$$= -3.7\,692 \qquad (4.38)$$

Substituting these back into the original equations (i.e. Eqns. 4.29, 4.30 and 4.31) gives

$$5 = 12.3\,846 + 0.154 - 7.5\,384 = 5.002 \tag{4.39}$$

$$10 = 6.1\,923 + 0.0\,385 + 3.7\,692 = 10 \tag{4.40}$$

$$1 = 12.3\,846 - 0.077 - 11.3\,076 = 1 \tag{4.41}$$

respectively. Note that small rounding errors have produced a slight discrepancy in expression 4.39.

Question 4.3 Solve the following system of equations for a, b and c:

$10.3 = 3a + 2b + c$
$7 = a + b + 1.3c$
$5 = 10a - 1.35b - 1.1c$

Check your answer by substituting your results back into the above equations.

4.4 Quality assurance

It is a good idea to consider ways that answers can be checked. It is so easy to make mistakes that methods for checking answers are immensely valuable. This section suggests three approaches and any one of them may, or may not, be useful for checking your mathematics in a particular case. The methods are:
1 Approximation: does the answer seem to be about the right size?
2 Dimensional analysis: does the answer have the right units?
3 Special cases: does the answer give sensible results in simple cases?
   Approximation is done by simplifying the numbers in a problem so that calculations can be performed quickly and without error. In the porosity versus depth problem where

$$\phi = \phi_0 \, e^{-z/\lambda} \tag{2.17}$$

we might have the case where $z = 4.2$ km, $\lambda = 1.9$ km and $\phi_0 = 0.92$. The porosity is then given by

$$\phi = 0.92 \, e^{-4.2/1.9}$$
$$= 0.26$$

To check this, make the approximations $z \approx 4$ km, $\lambda \approx 2$ km, $e \approx 3$ and $\phi_0 \approx 0.9$. This produces

$$\phi = 0.9 \times 3^{-2}$$
$$= 0.9/3^2 = 0.9/9$$
$$= 0.1$$

which is highly reliable since it does not require the error-prone use of a calculator. However, this is not very close to the 'exact' answer of 0.26 given above. Careful checking of the arithmetic shows an error was made in the 'exact' calculation. In fact $z = 2.4$ km was used instead of $z = 4.2$ km, a type of error very easily made on a calculator. Recalculation using the correct value for z yields a porosity of 0.101 which is much closer to the estimated value.

This agreement is unusually good since some of the values (λ and e) were increased when approximated whilst the others (z and ϕ_0) were decreased. It happens that, in this case, these alterations very nearly cancel one another exactly. This will not usually happen quite so well! The closeness of the approximations also depends strongly upon the function involved. Expressions involving exponentials generally approximate rather poorly whilst other functions (particularly straight lines for example) can give very good answers after approximation.

> **Question 4.4** Using approximation, check your answer to question 3.9 from the last chapter. (Hint: Use $a \approx -1 \times 10^{-4}$, $b \approx 1$, and approximate c as well. Do as much as possible without the use of a calculator.)

Dimensional analysis is a method for finding the correct units for an answer. The reason that this can be used to check mathematics is that, frequently, the units are known anyway and the result from a dimensional analysis should agree with the expected units. Thus, dimensional analysis allows you to check the form of an equation rather than the accuracy of a numerical answer (cf approximation). Take the Earth's average density example from Chapter 3. This was

$$\rho = 3g/4G\pi r \tag{3.13}$$

where the numerical values and units for the symbols involved are $g = 9.81$ m s^{-2}, $r = 6370$ km $= 6.37 \times 10^6$ m, $G = 6.672 \times 10^{-11}$ m^3 kg^{-1} s^{-2}. The result, which is a density, should be in units of kg m^{-3}. What are the units predicted by Eqn. 3.13? The procedure is to re-evaluate Eqn. 3.13 using the units of the symbols rather than their numerical values. Thus

$$\text{Units of } \rho = \frac{\text{Units of } 3 \times \text{Units of } g}{\text{Units of } 4 \times \text{Units of } G \times \text{Units of } \rho \times \text{Units of } r} \tag{4.42}$$

Numbers such as 3, 4 or π do not normally have units. These are called **dimensionless**. The units of the other terms are:
Units of g = m s^{-2}
Units of G = m^3 kg^{-1} s^{-2}
Units of r = m
Substituting these into Eqn. 4.42 gives

Units of ρ = m s^{-2}/[(m^3 kg^{-1} s^{-2})(m)]

Multiplying the m^3 on the bottom line by the m on the bottom line then gives

Units of ρ = m s^{-2}/(m^4 kg^{-1} s^{-2})

Finally, cancelling m s^{-2} from top and bottom yields

Units of ρ = 1/(m^3 kg^{-1})
            = kg m^{-3}

as expected. Thus, Eqn. 3.13 is dimensionally balanced. Equations which do not balance dimensionally must be wrong. Unfortunately, equations are not necessarily correct just because they balance; it is possible for an incorrect equation to balance perfectly.

Question 4.5 Rearrange

$$g = GM/r^2 \tag{3.6}$$

into an equation for G. Hence, using the fact that
Units of g = m s^{-2}
Units of M = kg
Units of r = m
show that
Units of G = m^3 kg^{-1} s^{-2}.

In more complex expressions such as

$$\text{Temperature} = az^4 + bz^3 + cz^2 + dz + e \tag{2.8}$$

all terms must have the same units. Thus, since the left-hand side of this equation has units of °C, the units of az^4, bz^3, cz^2, dz and e must also be °C. Now, the units of z are kilometres and therefore the units of the constants must be as follows

Units of a = °C km^{-4}
Units of b = °C km^{-3}
Units of c = °C km^{-2}
Units of d = °C km^{-1}
Units of e = °C

In many textbooks you will see a slightly different way of doing dimensional analyses. In this approach, the **dimensions** of a unit are expressed in terms of mass, length and time (abbreviated to M, L and T). For example, acceleration has units of m s^{-2} which is a length divided by a time squared. Thus, acceleration has dimensions LT^{-2}. If this procedure is repeated for all items in Eqn. 3.13, the result is

Dimensions of ρ

$$= \frac{\text{Dimensions of } 3 \times \text{Dimensions of } g}{\text{Dimensions of } 4 \times \text{Dimensions of } G \times \text{Dimensions of } \rho \times \text{Dimensions of } r}$$

$$= (LT^{-2}) / (L^3 M^{-1} T^{-2} L) = ML^{-3} \tag{4.43}$$

which balances since the dimensions of density are, indeed, mass divided by length cubed. In practice, it does not much matter which of these two approaches to dimensional analysis you use since they are entirely equivalent. However, I prefer the first in most situations since it is easier when quantities such as temperature are included and it also forces you to check that you have used consistent units for all quantities in the equation.

Special cases, the final checking method, is also useful for testing whether the general form for your answer is reasonable. Essentially, an expression is evaluated for a situation in which the correct answer is already known. In the radioactive dating method, for example, a specimen which formed very recently should still retain, more or less, the whole of its initial proportion of the radioactive element. Thus, in Eqn. 4.14

$$Q = Q_0 \exp(-t/\lambda) \tag{4.14}$$

we should obtain a value for Q equal to Q_0 when $t = 0$. Now, substituting $t = 0$ into this expression yields

$$\begin{aligned} Q &= Q_0 \exp(0) \\ &= Q_0 \times 1 \\ &= Q_0 \end{aligned}$$

as required. If, on the other hand, an equation of the form

$$Q = Q_0[1.0 - \exp(-t/\lambda)] \tag{4.44}$$

had been suggested for the way that Q reduces with time, this yields a value of

$$Q = 0.0$$

in the case of $t = 0$. Since this is not the answer we expect, Eqn. 4.44 must be wrong. N.B. As with dimensional analysis, it is possible for an incorrect answer to pass this test but a failure of the test means that the equation is definitely wrong (i.e. we can always say that an answer is definitely wrong but can never say that an answer is definitely right).

Question 4.6 Check that the equation for temperature, T, as a function of depth, z,

$$T = az + T_0$$

gives a sensible answer for the temperature at zero depth. In this expression, a is the temperature gradient and T_0 is the surface temperature.

4.5 Further questions

4.7 The thickness of a bottomset bed at the foot of a delta can often be well approximated by the expression

$$t = t_0 \exp(-x/X)$$

where t is thickness, x is distance from the bottomset bed start and t_0 and X are constants.
(i) Rearrange this expression into an equation for $\ln(t_0)$.
(ii) In a particular delta bottomset bed, the thicknesses are 5 and 0.1 m at $x = 1$ and 4 km, respectively. Substitute these values into your answer from (i) to give two equations. Hence, obtain and solve an expression for X.
(iii) Estimate the bottomset thickness at $x = 0$ km.

4.8 The temperature, T, in the Earth may be well approximated by an equation of the form

$$T = ar^4 + br^2 + c$$

where T is temperature, r is distance from the Earth's centre and a, b and c are constants. At distances from the centre of 1260, 5660 and 6260 km the temperatures are 4300, 1900 and 1150°C, respectively. Use these figures to determine suitable values for a, b and c. Does the resultant expression give sensible values for temperature at the Earth's centre and at the Earth's surface?

4.9 What are the units of a, b and c in question 4.8?

4.10 Based upon dimensional analysis and special cases, which of the following equations, for sediment ages in a lake bed, are definitely wrong?
(i) Age = (Depth × Rate) + Age of top
(ii) Age = (Rate/Depth) + Age of top
(iii) Age = (Depth/Rate) + Depth of top
(iv) Age = (Depth/Rate) + Age of bottom bed
where all ages are in ky, rate is in m ky^{-1} and all depths are in metres.

4.11 Use the spreadsheet *Simul.xls* to check my solution to Eqns. 4.17 and 4.18. Also use it to check the answers to questions 4.3 and 4.8.

4.12 Use *Simul.xls* to find the coefficients of

$$T = a_0 + a_2 r^2 + a_4 r^4 + a_6 r^6$$

where temperatures, T, and radius's, r, are given in the table below

Radius (km)	Temperature (°C)
0	4300
1260	4300
3560	3700
6260	1150

Use *Poly.xls* to calculate, and plot, the resulting function. Compare your results to those from question 4.8.

5 Trigonometry

5.1 Introduction

Trigonometry is the study of triangles. Triangles rather than, say, squares or hexagons because any other **polygon** (a closed shape with straight edges) can be constructed by adding triangles together (Fig. 5.1). Thus, if the properties of triangles are understood, any other polygon can also be dealt with.

Triangles are ideal for purposes such as mapping since there are simple rules relating the lengths of their sides to the size of their angles. Figure 5.2 illustrates the quantities which define a given triangle. This triangle has three sides of length a, b and c and three angles of size A, B and C. Note that length a is opposite angle A, b is opposite B and c is opposite C.

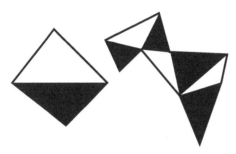

Fig. 5.1 Any polygon can be constructed from a set of triangles.

Fig. 5.2 A triangle is described by the length of its three sides and the size of its three angles.

Question 5.1 Using a ruler and protractor, sketch the following triangles and determine the unknown three quantities:
(i) $A = 20°$, $C = 100°$, $a = 4$ cm;
(ii) $C = 20°$, $a = 3$ cm, $b = 5$ cm.

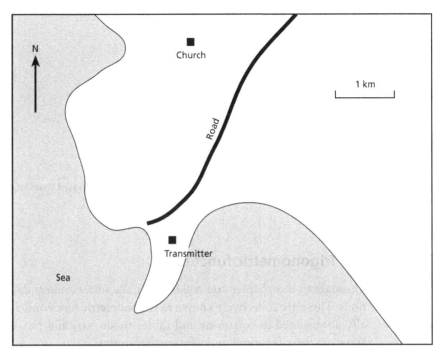

Fig. 5.3 Locate the exposure given the information in Question 5.2.

> **Question 5.2** Examine the map in Fig. 5.3 and measure the distance from the church to the transmitter. If, from an exposure, the church is seen to be located 45° west of north whilst the transmitter is due west, where is the exposure? How far is the exposure from the church and how far from the transmitter?

Angles, in geology, are normally measured in degrees since this is a convenient unit for measuring dips, strikes and other similar quantities. However, there are other units which can be used of which **radians** are the most important. An angle of one radian is about 57.3°. This may seem a very peculiar, and rather large, unit but there are good reasons for its use one of which will be explained in Chapter 8. For now, I will only point out that the radian is defined such that one complete rotation (i.e. an angle of 360°) is 2π radians (~6.28 radians).

> **Question 5.3** Given that 360° is equivalent to 2π radians, what are the following angles in radians? (Hint: What fraction of a complete rotation are these angles?)
>
> (i) 180°; (ii) 90°; (iii) 270°; (iv) 100°.

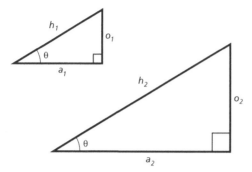

Fig. 5.4 Two similar triangles. The lower triangle has sides which are k times longer than those of the upper triangle. However, all angles are identical.

5.2 Trigonometric functions

Throughout this chapter you will be using the sine, cosine and tangent functions. These are collectively known as **trigonometric functions**. They are usually abbreviated in equations and tables to sin, cos and tan, respectively. What are these functions and why are they useful?

Figure 5.4 shows two **right-angled triangles** (triangles in which one angle is 90°), each of which contains the same angle, θ. However, the second triangle has sides which are k times longer than those of the first, i.e. $o_2 = k.o_1$, $a_2 = k.a_1$ and $h_2 = k.h_1$ where k is a constant. Triangles such as these, which are exactly the same shape but which are of different sizes, are known as **similar** triangles. Incidentally, I have denoted the lengths of the sides using o because this is the side opposite the given angle, a because this is the side adjacent to θ and h for hypotenuse which is the side opposite the right angle.

Now, for the larger triangle, the length of the opposite side divided by the length of the adjacent side is

$$o_2/a_2 = (k.o_1)/(k.a_1)$$
$$= o_1/a_1 \tag{5.1}$$

i.e. dividing the length of the opposite side by the length of the adjacent side gives the same value for both triangles. This value will only depend upon the angle θ. This ratio is called the **tangent** of θ (or tan(θ)) and can be found either by looking it up in tables or by the use of a calculator. Thus,

$$\tan(\theta) = \text{length of the opposite side / length of the adjacent side} \tag{5.2}$$

It is worth having a look at the ratios formed from pairs of sides other than a and o in Fig. 5.4. For example, the length of the opposite side divided by the length of the hypotenuse is also the same for both triangles since

$o_2/h_2 = (k.o_1)/(k.h_1)$
$\qquad = o_1/h_1$ (5.3)

This ratio, again, only depends upon the angle θ and is called the sine of θ or,

$\sin(\theta) =$ length of the opposite side/length of the hypotenuse   (5.4)

Finally, the length of the adjacent side divided by the length of the hypotenuse is a constant for the two triangles since

$a_2/h_2 = (k.a_1)/(k.h_1)$
$\qquad = a_1/h_1$ (5.5)

This ratio is called the cosine of θ, i.e.

$\cos(\theta) =$ length of the adjacent side / length of the hypotenuse   (5.6)

N.B. The definitions of tan, cos and sin given above are only true for right-angled triangles.

Question 5.4 The hypotenuse of a right-angled triangle is twice the length of one of the other sides. Calculate cos, sin and tan for the angles in the triangle. (Hint: Let one side have a length x giving a hypotenuse of length $2x$. Then use Pythagoras' theorem (i.e. $h^2 = a^2 + o^2$) to find the length of the third side. You will probably find a sketch helpful.)

What are these functions used for? Figure 5.5 illustrates a common situation in which the sine function can be used. The geological map (Fig. 5.5a) shows an alternating sequence of sandstone and limestone formations. One of the sandstone formations has an outcrop width of 1.25 km and its beds dip at 27°. What is the true thickness of this formation? Figure 5.5b shows how the apparent width, W, of a bed or formation is related to its true thickness, T, and its dip. From the definition of sine (Eqn. 5.4) it follows that

$\sin(\text{Dip}) = T/W$ (5.7)

which, after rearrangement, yields

$T = W \sin(\text{Dip})$ (5.8)

Substituting the known values for W and dip and using a calculator (or tables) to calculate $\sin(\text{Dip})$, gives

$T = 1.25 \sin(27°)$
$\quad = 1.25 \times 0.454$
$\quad = 0.567$ km
$\quad = 567$ m   (5.9)

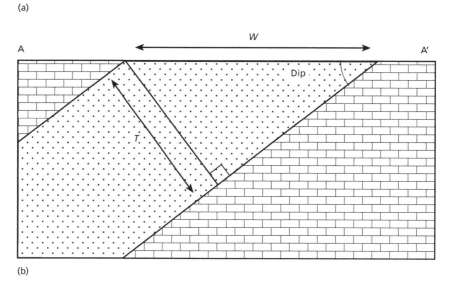

(a)

(b)

Fig. 5.5 (a) Geological map showing alternating sandstone and limestone bedding. One of the sandstone formations has a width of 1.25 km and a dip of 27°. (b) Vertical cross-section through a dipping bed which has a true thickness T and an apparent thickness on the surface of W.

Question 5.5 A cliff has a height of 130 m. A particular sedimentary bed outcrops at the cliff top and dips at 42.5° in a direction parallel to the cliff edge. Draw a sketch of this and, by considering the definition of the tangent function, determine how far away, horizontally, the same bed outcrops at the cliff base.

The **inverse trigonometric functions** produce the angle corresponding to a particular value for a sine, cosine or tangent. For example, $\sin(37°) = 0.602$ and therefore the inverse sine of 0.602 equals 37°. The inverse tangent, sine and cosine functions are sometimes called the arctangent, arcsine and arcosine functions. In equations they are denoted by $\tan^{-1}$, $\sin^{-1}$ and $\cos^{-1}$, respectively (e.g. $\sin^{-1}(0.602) = 37°$). Some calculators use a notation of atan, asin and acos instead or, very occasionally, arctan, arcsin and arcos.

The standard notation is very poor since there is a very similar notation for denoting powers of trigonometric functions. For example, the square of $\tan(\theta)$ (i.e. $\tan(\theta).\tan(\theta)$) is usually written $\tan^2(\theta)$. Thus, $\tan^{-1}(\theta)$ might be thought, erroneously, to be the same as $1/\tan(\theta)$. Unfortunately, this way of denoting the inverse trigonometric functions is very well established and is unlikely to be dropped now.

The fact that, with these inverse functions, angles can now be found from knowledge of their sines, cosines or tangents greatly increases the power of trigonometry. For example, the inverse tangent function can be used to determine true bed dips from a cross section which has vertical exaggeration. Geological cross sections frequently have different scales in the vertical and horizontal directions since data may be mapped over several kilometres horizontally but only extrapolated downwards for a few hundred metres. Figure 5.6 shows an example in which the cross section is 4 or 5 km wide but only about 100 m deep. The vertical exaggeration here is about 12 to 1 (i.e. the vertical scale is stretched 12-fold relative to the horizontal scale). Thus, to

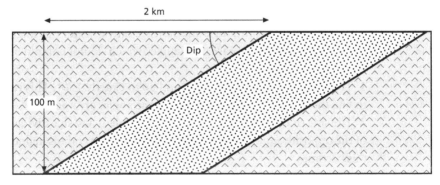

Fig. 5.6 Cross-section in which the vertical scale is about 12 times larger than the horizontal scale. The result is an apparent bed dip which is much greater than the true dip.

get a section in true scale, all vertical distances should be shrunk by a factor of 12. As a result, the beds, which appear to have a dip of about 30°, have a true dip which is much less. The true dip may be found by noting that, from Eqn. 5.2 and Fig. 5.6,

$$\begin{aligned} \tan(\text{dip}) &= \text{opposite/adjacent} \\ &= 100 \text{ m/2 km} = 100/2000 \\ &= 0.05 \end{aligned} \tag{5.10}$$

therefore, using the inverse tangent,

$$\begin{aligned} \text{dip} &= \tan^{-1}(0.05) \\ &= 2.86° \end{aligned} \tag{5.11}$$

Thus, the true dip is less than 3°, i.e. about one-tenth of the apparent dip in Fig. 5.6.

Question 5.6 A cliff has a height of 45 m. A bed at the cliff top outcrops 110 m, horizontally, from where it outcrops at the cliff base. Draw a sketch and determine the value of the tangent of the bed dip. Using the inverse tangent, find the dip in degrees.

5.3 Determining unknown angles and distances

In questions 5.1 and 5.2 earlier in this chapter, the angles and side lengths of several triangles were determined by drawing a sketch using the supplied information and measuring the unknown lengths and angles. Clearly, it would be more convenient and more accurate if the unknowns could be calculated, rather than measured, and this is indeed possible. In fact, the problems and examples discussed in the previous section have been doing precisely this for the special case of triangles containing a right angle. In question 5.6, for example, you were given the lengths of two sides and the value of one angle (i.e. 90°) from which the remaining two angles and one side length could be calculated (although question 5.6 only asks for one angle).

For the more general case of triangles which do not contain a right angle, three rules are needed:

1 The 180° rule. The angles must add up to exactly 180°. This rule allows us to find the third angle whenever two of the angles are known.

2 The sine rule. For a given triangle, the length of any side divided by the sine of the opposite angle is a constant. In terms of the symbols defined in Fig. 5.2 this becomes

$$\frac{a}{\sin(A)} = \frac{b}{\sin(B)} = \frac{c}{\sin(C)} \tag{5.12}$$

3 The cosine rule. This is a generalization of Pythagoras' theorem to cover non-right-angled triangles. In terms of the symbols defined in Fig. 5.2, the cosine rule is

$$a^2 = b^2 + c^2 - 2bc.\cos(A) \qquad (5.13)$$

or

$$b^2 = a^2 + c^2 - 2ac.\cos(B) \qquad (5.14)$$

or

$$c^2 = a^2 + b^2 - 2ab.\cos(C) \qquad (5.15)$$

Question 5.7 Using the symbols from Fig. 5.2:
(i) If $A = B = 70°$, use the 180° rule to find angle C,
(ii) $b = 3$ km, $c = 2$ km and $C = 40°$, use the sine rule to find angle B,
(iii) If $b = 3$ km, $c = 1$ km and $A = 37°$, use the cosine rule to find length a.

Question 5.8 If angle A is a right angle, show that Eqn. 5.13 reduces to Pythagoras' theorem $a^2 = b^2 + c^2$.

The 180° rule, the sine rule and the cosine rule are used in different ways and different orders depending upon the information known at the start of a specific problem. In general, a triangle is characterized by six quantities (i.e., three lengths and three angles) and all six can be found provided at least one length is known plus any two other pieces of information.

Given the three rules, and three pieces of information, most problems can be solved in several different ways. Suppose, for example, that three sides and zero angles are known. The first step is to use the known lengths to calculate one of the unknown angles. This implies that the cosine rule should be used since the other rules all involve more than one angle. Equation 5.13 can be rearranged into

$$\cos(A) = \frac{b^2 + c^2 + a^2}{2bc} \qquad (5.16)$$

which can then be used to find angle A by using the inverse cosine function to give

$$A = \cos^{-1}[(b^2 + c^2 - a^2)/2bc] \qquad (5.17)$$

Three sides and one angle, A, are now known. At this point it is possible to proceed using either the cosine rule or the sine rule. However, where possible and for reasons given later, it is usually best to avoid using the sine rule. Hence, rearranging Eqn. 5.14 gives

$$\cos(B) = \frac{a^2 + c^2 - b^2}{2ac} \tag{5.18}$$

and then, using the inverse cosine function

$$B = \cos^{-1}[(a^2 + c^2 - b^2)/2ac] \tag{5.19}$$

The final unknown quantity is the angle C which, using the $180°$ rule, is found from

$$C = 180 - A - B \tag{5.20}$$

Question 5.9 Two exposures, 500 m apart, are 400 m and 200 m, respectively, from a church. Calculate, using Eqns. 5.17, 5.19 and 5.20, the angles contained by the triangle defined by the two exposures and the church.

Question 5.10 Find the unknown quantities in the following:
(i) $A = 40°, b = 5$ km, $c = 2$ km;
(ii) $A = 40°, b = 3$ km, $a = 2$ km;
(iii) $A = 40°, B = 60°, a = 3$ km.

5.4 Cartesian coordinates and trigonometric functions of angles bigger than 90°

The definitions of the trigonometric functions given in Section 5.2 are only valid for angles less than $90°$. However, triangles frequently have one angle greater than this. So, how are these functions defined in these cases? It is easiest to begin by first discussing **Cartesian coordinates** (Fig. 5.7). This is a way of specifying any location in a plane by giving the horizontal and vertical distance from an **origin** (the centre of the coordinate system where $x = y = 0$). Point A, for example, is at the location $x = 5$, $y = 10$. This is frequently abbreviated to 'the point (5,10)'. Note that points to the left of the origin have a negative x coordinate and points below the origin have a negative y coordinate.

This same coordinate system could be used to specify angles by drawing lines between the origin and these points (Fig. 5.8). The corresponding angle is that formed between the x-axis and the line, measured in an anticlockwise sense. (N.B. When measuring compass bearings, angles are measured clockwise around from North. It is unfortunate that mathematicians and cartographers have settled on different conventions but you will have to get used to using these in different contexts.)

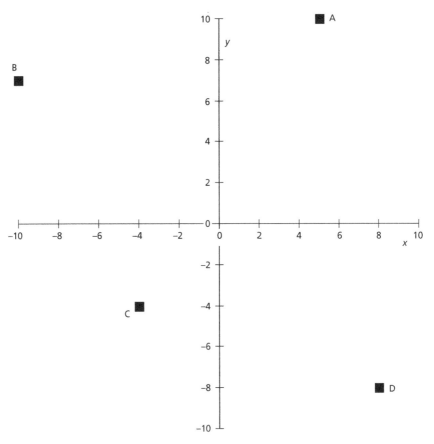

Fig. 5.7 Cartesian coordinates to specify locations of points. For example point A is at $x = 5$, $y = 10$ and point C is at $x = -4$, $y = -4$.

Starting with the point A: the angle, θ_a, corresponding to the point $(5,10)$ has a tangent of

$$\begin{aligned}
\tan(\theta_a) &= \text{opposite/adjacent} \\
&= y\text{-coordinate}/x\text{-coordinate} \\
&= 10/5 \\
&= 2.0
\end{aligned} \tag{5.21}$$

i.e. the angle is $\tan^{-1}(2) = 63.4°$. Now, the same procedure could be attempted with point B at $(-10,7)$. The angle of interest is now between $90°$ and $180°$. Note that x is therefore negative. In other words, the length of the adjacent side is negative giving

$$\begin{aligned}
\tan(\theta_b) &= \text{opposite/adjacent} \\
&= 7/-10 \\
&= -0.7
\end{aligned} \tag{5.22}$$

Thus, for this case the tangent is a negative number.

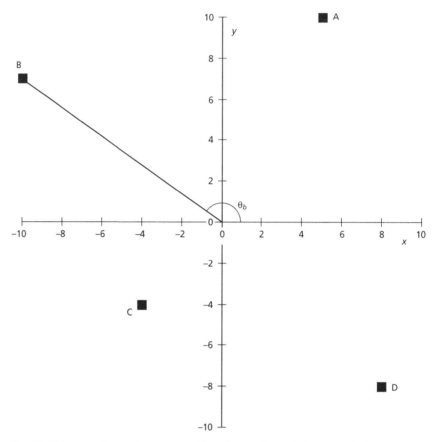

Fig. 5.8 Using cartesian coodinates to specify angles. Angles are defined anti-clockwise from the *x*-axis to each of the lines. For clarity, only one angle is shown.

Question 5.11 Repeat the above procedure to find the tangents of the angles produced using points C and D in Fig. 5.8.

The same process could now be attempted for the cosines and sines of the angles produced by points A, B, C and D. For the sine and cosine calculations, the hypotenuse is the line from the origin to the point and is always taken to have a positive length. Figure 5.9 summarizes the results by displaying sine, cosine and tangent for angles between 0° and 360°.

Note that there is more than one angle which gives rise to any particular value for the sine, cosine or tangent (e.g. $\tan(45°) = \tan(225°) = 1.0$). For this reason, the angles obtained from calculating the inverse trigonometric functions are not unique. Thus, your calculator would give $\tan^{-1}(1.0) = 45°$ but the answer could be 225°. In general, there are two angles between 0° and 360° which give rise to any given value for sine, cosine or tangent. You

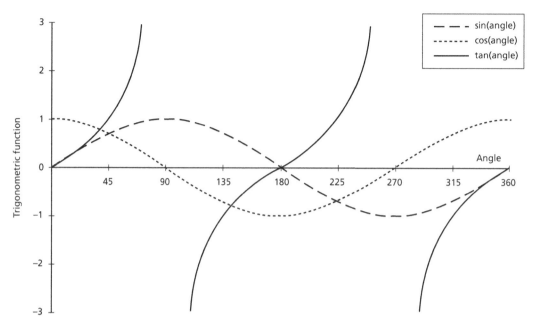

Fig. 5.9 The sine, cosine and tangent functions for angles between 0° and 360°.

Fig. 5.10 What size is θ in this triangle?

should be aware that this non-uniqueness can, occasionally, result in incorrect answers. Consider Fig. 5.10 in which a triangle is shown with an unknown angle, θ, clearly much larger than 90°.

The obvious way to determine θ is to use the sine rule which leads to

$$\sin(20)/5 = \sin(\theta)/10 \qquad (5.23)$$

thus

$$\sin(\theta) = 2 \sin(20)$$
$$= 0.684 \qquad (5.24)$$

giving

$$\theta = \sin^{-1}(0.684)$$
$$= 43.2° \qquad (5.25)$$

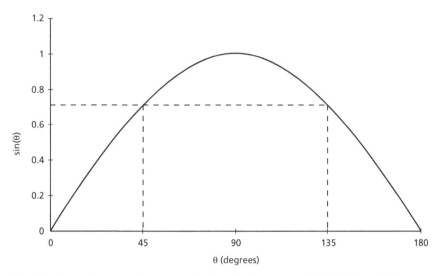

Fig. 5.11 Non-uniqueness of the inverse sine function. For Example sin⁻¹(0.684) = 43.2° or 136.8° (i.e. 180–43.2).

Fig. 5.12 Compare this to Fig. 5.10. The information specified is precisely the same but, this time, a solution less than 90° is plausible.

which is plainly wrong. The reason for this is that there are two angles between 0° and 180° which have a sine of 0.684. Inspecting Fig. 5.11, an angle of 180 – 43.2 degrees has the same sine as 43.2°. Thus, the correct answer is 136.8°.

The reason for this uncertainty is simply that the other answer, 43.2°, is possible with the information given. This is illustrated by Fig. 5.12 which has exactly the same known starting information (2 sides and one opposite angle) but which actually has a solution of 43.2°. In the case of two sides and one opposite angle, we must also know if the other opposite angle is less than or greater than 90°.

In conclusion, whenever you use the inverse sine, cosine or tangent functions, calculate both solutions and then decide which is appropriate in the particular case you are investigating. Note that, for the inverse cosine and inverse tangent functions, the larger of the two solutions is greater than 180° and, therefore, can be ignored if the answer is the angle of a triangle.

However, in some other types of problem, not discussed in this book, these large solutions may be valid. The two solutions for the inverse sine function, on the other hand, are both less than 180° and therefore are both plausible. This is the reason that, earlier, I suggested avoiding the sine rule whenever possible.

5.5 Trigonometry in a three-dimensional world

Up to now, I have used trigonometry purely in two dimensions. However, geology is a three-dimensional subject. Figure 5.13a illustrates a typical three-dimensional problem. Imagine a dipping bedding plane outcropping on a cliff face which is not parallel to the direction of maximum slope. This will result in an apparent dip, on the cliff face, which is less than the true dip. The most extreme case is where the cliff face is at right angles to the direction of dip (i.e. the cliff is in the *strike* direction). In this extreme case the apparent dip of the beds is zero! Is there a simple relationship between the true dip and the apparent dip?

Figure 5.13b shows a construction for determining this relationship. In this diagram the bedding plane has a true dip, θ, in a direction parallel to the x-axis. The line OC represents the direction along which the bedding plane is cut (i.e. the cliff face). This direction is at an angle α to the dip direction. This results in an apparent dip of θ'. Note that the angles COA, BOA and OBC are all right angles. From this it follows that

$$\tan(\theta) = OA/OB \tag{5.26}$$

$$\tan(\theta') = OA/OC \tag{5.27}$$

and

$$\cos(\alpha) = OB/OC \tag{5.28}$$

Equation 5.27 and 5.28 can be combined as follows

$$\tan(\theta') = OA/OC = (OA/OB).(OB/OC)$$
$$= (OA/OB) \cos(\alpha) \tag{5.29}$$

which, together with Eqn. 5.26, leads to

$$\tan(\theta') = \tan(\theta) \cos(\alpha) \tag{5.30}$$

which relates the apparent dip, θ', to the true dip, θ, and the angle, α, which the cliff makes to the dip direction. From this, the apparent dip can be found using

$$\theta' = \tan^{-1}[\tan(\theta) \cos(\alpha)] \tag{5.31}$$

or the true dip can be calculated from the apparent dip using

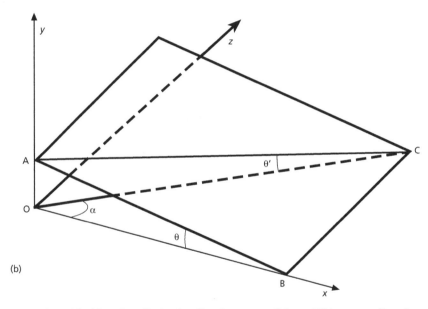

(a)

(b)

Fig. 5.13 (a) A bedding plane dipping in a direction not parallel to a cliff (represented here by the back face of the cuboid). This results in an apparent dip on the cliff face which is less than the true dip. (b) Construction for determining the true dip from the apparent dip. Triangle AOB is parallel to the true dip direction whilst AOC is parallel to the cliff face.

$$\theta = \tan^{-1}[\tan(\theta')/\cos(\alpha)] \tag{5.32}$$

If the apparent dip is 32°, measured in a direction 25° from the direction of maximum dip, the true dip is

$$\begin{aligned}\theta &= \tan^{-1}[\tan(32)/\cos(25)]\\ &= \tan^{-1}[0.625/0.906]\\ &= \tan^{-1}[0.689]\\ &= 34.6° \end{aligned}\tag{5.33}$$

> **Question 5.12** On a cliff face, the apparent dip is 25° whilst the true dip is 35°. What is the angle between the cliff face and the strike direction?

Finding the true dip and its direction, in the field, can become even more complex than indicated. Additional difficulties not considered so far are such things as the effect of uneven topography, non-planar bedding and measurements made on inclined surfaces. Fortunately, there is a much easier approach using **stereographic projection**, a subject that will be introduced in the next chapter.

5.6 Introduction to vectors

This section will give a brief introduction to **vectors**. A more detailed treatment is beyond the scope of this book, but the most common vector operation (**vector addition**) is covered.

A vector is any quantity which has a direction as well as a magnitude. River channels, for example, can be described in terms of their direction and rate of flow. Another example is the Earth's magnetic field which, at any given point on the Earth's surface, has a definite direction (roughly speaking the field points North with a dip which depends upon latitude) as well as a definite strength (the strength increases towards the poles). Quantities which only have magnitude but no direction are called **scalars** (e.g. temperature).

> **Question 5.13** Are the following vector or scalar quantities?
> (i)   Mass;
> (ii) Gravitational acceleration;
> (iii) Age;
> (iv) The line joining an exposure location to a church.

Figure 5.14 shows a series of vectors representing flow at various locations on a river. The arrows point in the flow direction and have a length which is proportional to the flow speed. These arrows are diagrammatic representations of the flow vectors. These vectors can be denoted by the letters $\underline{A}$, $\underline{B}$, $\underline{C}$ and $\underline{D}$ where underlining is a way of indicating that they are vectors. An alternative notation is to indicate vectors by using boldface (i.e. **A**, **B**, **C** and **D**). In this section, I shall deliberately alternate these so that you get used to seeing vectors written both ways. Obviously, if you are writing vector expressions by hand, it is easiest to use the underlining convention.

An important property of vectors is that they may be added together. Vector addition is simply the process of combining the vectors nose-to-tail

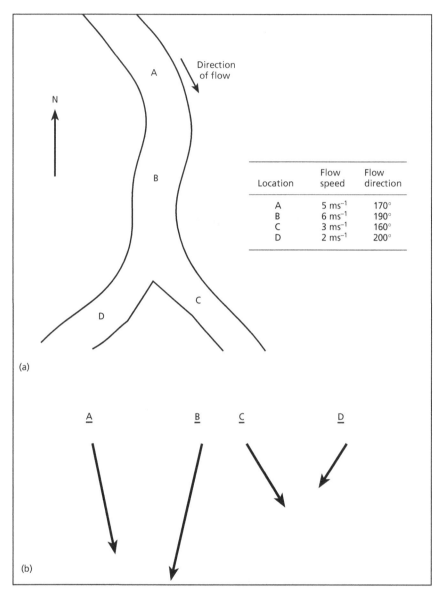

Location	Flow speed	Flow direction
A	5 ms⁻¹	170°
B	6 ms⁻¹	190°
C	3 ms⁻¹	160°
D	2 ms⁻¹	200°

(a)

(b)

Fig. 5.14 (a) A river flowing, roughly, southwards. At points A, B, C and D the river speed and direction are as shown in the table. (b) Vector representation of the river flow at A, B, C and D. The direction of the arrows shows the flow direction and their length is proportional to the speed.

as shown in Fig. 5.15. The resultant vector is obtained by drawing a vector from the tail of the first vector to the nose of the last. This operation can be algebraically represented by the equation

$$\underline{r} = \underline{a} + \underline{b} + \underline{c} + \underline{d} \tag{5.34}$$

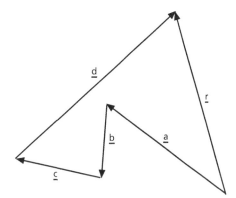

Fig. 5.15 Vector addition. Vectors $\underline{a}$, $\underline{b}$, $\underline{c}$ and $\underline{d}$ are added nose-to-tail as shown to give the result $\underline{r}$.

An important point about this addition is that it is not the same as adding the vector lengths and vector directions separately.

Question 5.14 Draw a set of x–y axes on a sheet of paper. Then draw the vectors:
(i) Vector $\underline{a}$: Length 3 cm, direction 10° clockwise from the x-axis;
(ii) Vector $\underline{b}$: Length 5 cm, direction 50° clockwise from the x-axis;
(iii) Vector $\underline{c}$: Length 3 cm, direction 190° clockwise from the x-axis;
(iv) The vector $\underline{d} = \underline{a} + \underline{b}$;
(v) The vector $\underline{e} = \underline{a} + \underline{c}$.

Another useful operation on vectors is **scalar multiplication** (don't confuse this with the **scalar product,** a more sophisticated vector operation, which will not be discussed further here). In scalar multiplication the vector length is simply increased by multiplying it by a scalar. Thus, a vector in the direction 13° E of N with a magnitude of 5 km becomes, after scalar multiplication by 3, a vector in the same direction (i.e. 13° E of N) but with a magnitude increased to 15 km. A small complication is the effect of multiplying by a negative quantity (e.g. −3). In this case, the direction of the vector is reversed and so multiplying a vector in the direction 13° E of N with a magnitude of 5 km becomes, after scalar multiplication by −1, a vector in the direction 193° E of N with a magnitude of 5 km.

Vector addition and scalar multiplication together allow a new way of specifying a vector to be introduced. The idea is to specify the vector in terms of the lengths of **component vectors** in the x and y directions. This is illustrated in Fig. 5.16. The x-component and the y-component sum to produce the given vector. The vector can then be written down as

$$a = x\mathbf{i} + y\mathbf{j} \tag{5.35}$$

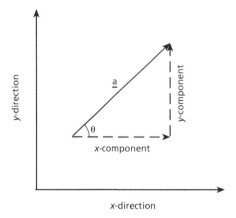

Fig. 5.16 Any vector can be thought of as the result of summing an x-component vector and a y-component vector.

where i and j are vectors of length 1.0 in the x and y directions, respectively, and where x and y are the lengths of the x and y components. Thus, xi is a vector of length x in the x-direction and yj is a vector of length y in the y-direction. In other words, vector a is the sum of the x and y component vectors. Vectors such as i and j, which are of unit length, are known as **unit vectors**.

To convert one form of vector specification into the other, it is only necessary to use a little trigonometry. From Fig. 5.16, the lengths of the x and y components are

$$x = a \cos(\theta) \tag{5.36}$$

and

$$y = a \sin(\theta) \tag{5.37}$$

where a is the length of vector a and θ is the angle which vector a makes to the x-direction. Thus, if the length, a, and direction, θ, of the vector are known, the x and y components can be easily found. To convert from components to vector magnitude, Pythagoras' theorem gives

$$\text{vector length, } a = \sqrt{x^2 + y^2} \tag{5.38}$$

The vector direction follows from the definition of the tangent function and, for the case of Fig. 5.16, gives

$$\text{vector direction, } \theta = \tan^{-1}(y/x) \tag{5.39}$$

Question 5.15 Use the definitions of the sine and cosine functions and Fig. 5.16 to derive Eqns. 5.36 and 5.37 above.

Fig. 5.17 Slip vectors for two faults. The overall slip of block 1 relative to block 3 can be obtained by vector addition of these two slip vectors.

The advantage of recasting vectors in terms of their components is that it makes vector addition much simpler. To add vectors together, you simply add the components. The example illustrated in Fig. 5.17 should make everything much clearer. In this example there are two closely spaced faults with different throws and different fault dips. The question is: what is the total movement of block 1 relative to block 3? Using vectors makes this problem extremely straightforward. All we do is add the slip vector (i.e. a vector representing the direction of slip and amount of throw) for fault 1 to the slip vector for fault 2. So, what are the slip vectors, s_1 and s_2, for each of the faults? Using trigonometry in an identical manner to that used for determining Eqns. 5.36 and 5.37 leads to

$$s_1 = 18.8 \cos(46°)i + 18.8 \sin(46°)k$$
$$= 13.1i + 13.5k \tag{5.40}$$

and

$$s_2 = 13.5 \cos(38°)i + 13.5 \sin(38°)k$$
$$= 10.6i + 8.3k \tag{5.41}$$

giving a resultant, total slip, of

$$s = s_1 + s_2$$
$$= (13.1 + 10.6)i + (13.5 + 8.3)k$$
$$= 23.7i + 21.8k \tag{5.42}$$

However, we would probably wish to have the final answer in the form of the dip and throw of a single fault which would give the same effect. Thus, we need to convert back from vector components to vector direction and magnitude. Applying the same principles as used in Eqns. 5.38 and 5.39 to the fault throw problem gives

$$\text{Total throw} = \sqrt{23.7^2 + 21.8^2} = 32.2 \text{ m} \tag{5.43}$$

and

$$\text{Equivalent single fault dip} = \tan^{-1}(21.8/23.7)$$
$$= 42.6° \tag{5.44}$$

Thus, a single fault dipping at 42.6° with a throw of 32.2 m, would have given an identical vertical and horizontal movement to block 1 relative to block 3. The resultant dip direction is called a **vector mean** direction. This kind of analysis would be useful when examining the variation in extension across fault systems as we move along strike. The number, dip and throw of faults generally varies along strike but by summing the individual slip vectors we could directly compare the size and direction of extension. In general, this analysis would need to be performed as a problem in three dimensions, in which case the slip vectors have three components rather than two. However, apart from this, the methods used would be identical.

Question 5.16 Three adjacent faults have throws and dips of:
(i) 10 m at 60°; (ii) 5 m at 65°; (iii) 12 m at 45°. Calculate the total slip vector.

5.7 Further questions

5.17 Evaluate:
(i) $\cos(15°)$; (ii) $\sin(1.2 \text{ radians})$; (iii) $\tan^{-1}(0.5)$; (iv) $\cos^2(27°)$;
(v) $(\tan(0.5°))^{-1}$

5.18 If the Earth were a perfect sphere, show that the radius, r, of a circle of latitude is given by

$$r = R \cos(\phi)$$

where R is the Earth's radius and ϕ is the latitude.

5.19 An alluvial fan slopes at an angle of 5° to the horizontal and the distance from the fan origin to its base (measured along the fan surface) is 5 km. Calculate the height of the fan origin above its base.

5.20 Look at Fig. 5.18 which shows a geological map (Fig. 5.18a) and a section (Fig. 5.18b) drawn from a cliff at the location shown on the map. The map indicates that in the area of the section the direction of strike is 72° E of

(a)

(b)

Fig. 5.18 (a) Geological map showing the position and alignment of the section shown in (b) and the strike direction of the beds.

N and the section itself is aligned along 130° E of N. From the information shown on the section, determine the true dip of the beds.

5.21 Remanent magnetism of 10 specimens collected from a Tertiary sill had the following azimuthal directions in degrees E of N:

331, 5, 347, 351, 3, 342, 338, 355, 349, 17.

Assuming that the remanent magnetism strengths are equal (say unity), calculate the vector mean direction for these measurements. Do this either graphically or by calculating vector components.

5.22 Carbonate platform foreslopes can be much steeper than those of deltas. If the water depth is 100 m only 500 m offshore from the slope top, what is the slope?

5.23 A river plume is transporting suspended sediment southwards at a rate of 0.1 m s^{-1}. However, a tidal current of 0.5 m s^{-1} moving in a direction 60° E of N and a longshore drift moving west at 0.2 m s^{-1} are superimposed on this. Using vectors, calculate the resultant drift rate and direction of the plume.

5.24 Use the spreadsheet *Trig.xls* to check the answers to questions 5.1, 5.2, 5.7, 5.9 and 5.10.

5.25 Use the spreadsheet *Vsum.xls* to check the answers to questions 5.16 and 5.21.

6 More about graphs

6.1 Introduction

In one sense geologists frequently have too much data. A field geologist may have a notebook full of dip, strike and location measurements, a geochemist may have analyses of 10 different elements in 100 different rock samples, or a geophysicist may have more than a kilometre of computer tape for every kilometre of a 1000 km seismic survey. In all these cases the problem is the same, the scientist involved must somehow make sense of a mass of data that is far too large to be digested raw. There are three things that can be done about this.

(i) Throw away most of the data. Usually this means ignoring all data which does not fit some preconceived notion. This is very definitely not recommended although it is quite frequently done!

(ii) Perform a statistical analysis. This is the subject of the next chapter.

(iii) Plot the data on a graph which will allow the general properties of the data to be visualized. This is the subject of this chapter.

In fact, although I have separated them here, statistics and graphing are subjects which overlap very significantly.

This chapter deals with graphs in which each data item is plotted as a point on a suitable piece of graph paper. The most common graph of this type has already been used extensively, particularly in Chapter 2. This is the simple x–y graph which has two axes at right angles to each other, representing two different quantities. Figure 6.1, shows such a graph which plots sediment density against depth in a well. Each point represents a specific measurement of depth and density. The remainder of this chapter is about variations upon this simple theme.

6.2 Log-normal and log-log graphs

The use of logarithms, to enable a wide spread of data to be visualized, has already been introduced in Chapter 2. Table 6.1 gives the masses of various modern and extinct animals together with the total areas of the soles of their feet (this is relevant to whether these animals could walk on soft mud without sinking in and can help to indicate the environment in which they lived). This data is plotted on an x–y type graph in Fig. 6.2. Note that all the points except

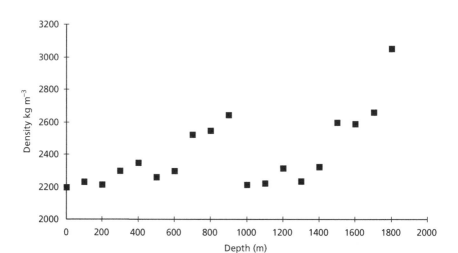

Fig. 6.1 A simple x–y plot of sediment density versus depth in a particular well.

Table 6.1 Masses and total foot area for modern and extinct animals. Data taken from Alexander, R. (1989). *Dynamics of Dinosaurs and other Extinct Giants*, Columbia University Press, New York.

Animal	Mass (kg)	Log(mass)	Foot area (m²)
Apatosaurus	35 000	4.54	1.2
Tyrannosaurus	7000	3.85	0.6
Iguanodon	5000	3.70	0.4
African Elephant	4500	3.65	0.6
Cow	600	2.78	0.04
Human	70	1.85	0.035

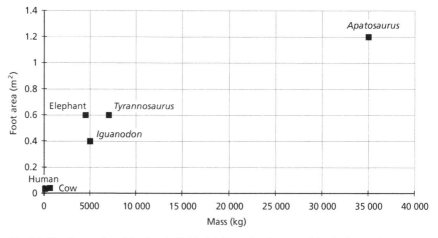

Fig. 6.2 Simple x–y plot of the data in Table 6.1. Note that five out of the six data points are squeezed into the leftmost fifth of the graph.

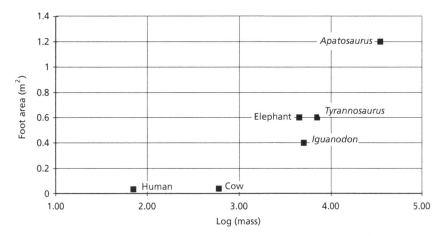

Fig. 6.3 Plot of foot area as a function of logarithm of the mass. Data is now much better spread across the plot.

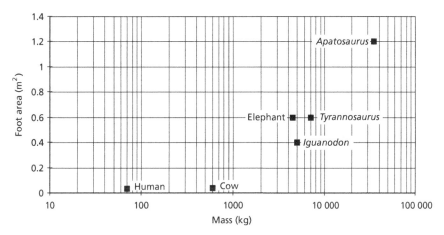

Fig. 6.4 An alternative to Fig. 6.3. Plot the raw data but use a logarithmically scaled axis. The resultant graph is the same shape as before but is easier to read.

one are squeezed into the leftmost fifth of this graph which makes the graph difficult to analyse. A solution to this problem is to plot using the logarithm of the mass instead as discussed in Section 2.8 and the result of doing this is shown in Fig. 6.3.

The problem with Fig. 6.3, however, is that it is now difficult to read off values on the horizontal axis. For example, without looking at Table 6.1, what is the mass of an elephant? You have to read down to the axis (gives 3.65) and then take the inverse logarithm (i.e. mass = $10^{3.65}$ = 4467 kg). This is rather tedious and error prone. An alternative, shown in Fig. 6.4, is to use a logarithmically scaled axis. Note that the distance on the horizontal axis between 100 kg and 1000 kg (a 10-fold increase) is the same as the distance between 10 kg and 100 kg (also a 10-fold increase). The result is a graph

Magnitude	Number per year
8	1
7	18
6	108
5	800
4	6 200
3	49 000
2	300 000

Table 6.2 Average earthquake frequency between 1918 and 1945. Data from Gutenberg, B. and Richter, C.F. (1954). *Seismicity of the Earth and Associated Phenomena*, Princeton University Press, Princeton.

whose shape is identical to that of Fig. 6.3 but from which it is much easier to read the mass of any given animal. Such a graph is known as a **log-normal** plot since one axis (the horizontal one in this case) is scaled logarithmically whilst the other has a normal scale. Note that the grid lines initially go up in multiples of 10 (i.e. the left-most grid line is for 10 kg, the second one for 20 kg etc.) until 100 kg is reached. Then the grid lines go up in multiples of 100 kg (i.e. the next grid line is for a mass of 200 kg) until 1000 kg is reached. Grid lines then increase in multiples of 1000 kg. Hence, the mass of the elephant can be read off the axis as around 4500 kg.

Question 6.1 Table 6.2 gives the average frequency (number per year) of earthquakes of various magnitudes over the period 1918 to 1945.
Using this data:
(i) Plot the frequency as a function of magnitude on normal graph paper. (Frequency should be on the vertical axis.)
(ii) Plot the frequency as a function of magnitude on log-normal graph paper.
(iii) Plot log(frequency) against magnitude on normal graph paper. Estimate the constant b in the equation $logN = k - bM$ where N is frequency, M is magnitude and k is a constant.
   It would have been more difficult, although not impossible, to estimate b from the second graph you plotted. Thus, if the main objective is to display the data more clearly, use log-normal graph paper but, if the objective is to estimate a parameter such as b, take logarithms first and plot on normal graph paper.

In Table 6.1, the foot areas are also spread over a rather large range. It might be useful, therefore, to take logarithms of the areas or to use logarithmically scaled axes in both directions. Figure 6.5 is an example of such a **log-log** plot.

Question 6.2 Using Fig. 6.5, what is the mass and total foot area of *Brachiosaurus*?

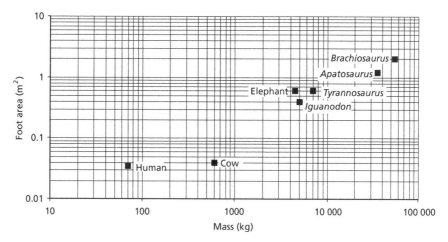

Fig. 6.5 The data from Table 6.1 plotted using logarithmic axes in both directions.

Incidentally, in this example, using a logarithmic scale for the vertical axis has only made a marginal improvement. However, in other cases it will make a much more useful alteration in the distribution of the data across the graph.

6.3 Triangular diagrams

Triangular diagrams can be used whenever you wish to visualize the relative proportions of three components making up a specimen. Common examples are:

(i) The proportions of sand (particles between 2 and 0.063 mm diameter), silt (0.063–0.004 mm) and clay (less than 0.004 mm) in a sedimentary rock;

(ii) An **AFM diagram** which shows the proportions of alkalis, iron and magnesium in a volcanic rock.

Figure 6.6 shows simple cases from the sedimentological example. Point A sits in the corner marked '100% clay' and represents a sediment containing only clay. Similarly, points B and C represent rocks containing exclusively

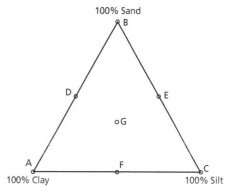

Fig. 6.6 A triangular diagram showing the proportion of clay, sand and silt for seven different sedimentary rock specimens.

sand and silt, respectively. Point D lies half way along a line joining 100% clay to 100% sand. It represents a sediment half of which is clay and half of which is sand. Similarly, point E is a 50 : 50 sand–silt mixture whilst point F is a 50 : 50 clay–silt mixture. Finally, point G is in the centre of the triangle, equidistant from all three edges, and represents a sediment which consists of a 1/3 clay, 1/3 sand and 1/3 silt mixture.

It is also, I think, fairly obvious where to plot a point corresponding to, say, 40% clay and 60% sand and which, therefore, contains no silt. This point will be on the line joining 100% sand to 100% clay and will be 40% of the distance along from sand to clay (or, equivalently, 60% of the distance along from clay to sand), i.e. slightly closer to sand than to clay.

Question 6.3 Plot the 40% clay, 60% sand point onto Fig. 6.6.

These examples are relatively straightforward and it is quite easy to see where on the triangular plot each of these points should go. What about a sediment containing 36% sand, 24% silt and 40% clay? Figure 6.7 illustrates how this

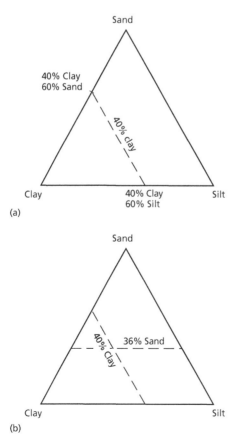

Fig. 6.7 Plotting a point which is 40% clay, 24% silt and 36% sand.

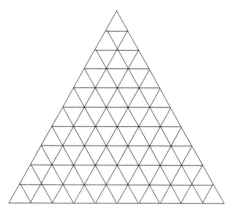

Fig. 6.8 A triangular diagram net. Lines are drawn here at 10% increments although most such graph paper would also have 1% intervals marked.

is done. In Fig. 6.7a, a line has been drawn which connects the 40% clay, 60% sand point to a point representing 40% clay, 60% silt. All points along this line contain 40% clay but have differing amounts of sand and silt making up the remaining 60%. Similarly, Fig. 6.7b shows a line representing all points which have 36% sand. The point where the 40% clay line intersects the 36% sand line is, of course, a point representing a sediment with 36% sand and 40% clay and which must, therefore, be 24% silt.

To assist in accurate plotting of such points, a triangular net similar to that shown in Fig. 6.8 is used. For clarity in this illustration, the lines are drawn at 10% intervals, although these lines will usually be plotted at 1% intervals on most sheets of triangular graph paper.

Question 6.4 Use Fig. 6.8 or some triangular graph paper to plot an AFM diagram as follows.

The left corner of the plot represents 100% ($Na_2O + K_2O$). The right-hand corner represents 100% MgO. The top corner represents 100% ($FeO + Fe_2O_3$). Mark these points on your graph and then plot the following data which is taken from a set of related volcanic rocks.

(i) 10% ($Na_2O + K_2O$), 45% MgO, 45% ($FeO + Fe_2O_3$).
(ii) 10% ($Na_2O + K_2O$), 35% MgO, 55% ($FeO + Fe_2O_3$).
(iii) 10% ($Na_2O + K_2O$), 25% MgO, 65% ($FeO + Fe_2O_3$).
(iv) 12% ($Na_2O + K_2O$), 20% MgO, 68% ($FeO + Fe_2O_3$).
(v) 15% ($Na_2O + K_2O$), 15% MgO, 70% ($FeO + Fe_2O_3$).
(vi) 18% ($Na_2O + K_2O$), 12% MgO, 70% ($FeO + Fe_2O_3$).
(vii) 23% ($Na_2O + K_2O$), 12% MgO, 65% ($FeO + Fe_2O_3$).

A graph such as this can furnish significant information about the evolution of a volcanic rock series. However, the way in which this is done, as well as the details of how to obtain the numbers to plot, is beyond the scope of this book.

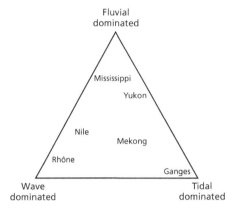

Fig. 6.9 Classification of delta types using a triangular diagram.

Before leaving the subject of triangular diagrams, it is worth mentioning that they can be used for classification of geological features if there are three clear **end members** to such a classification scheme. For example, deltas are commonly classified as being dominated by fluvial, wave or tidal processes. However, real deltas are influenced to some extent by all three types of process and will not be accurately represented by a simple threefold classification scheme. The solution is to use a triangular diagram to represent all possible deltas (Fig. 6.9). Real deltas will then fall at some point within the diagram which represents the proportions of fluvial, wave and tidal effects governing the geometry.

6.4 Polar graphs

Some types of data are naturally cyclic. For example, the data in Table 6.3 gives the strength of the non-dipole portion of the Earth's magnetic field at

Table 6.3 Non-dipole magnetic strength (in micro tesla) at various locations around the Earth's equator.

Longitude	Non-dipole strength (μT)
0	17.5
30	13
60	6
90	9.5
120	9.5
150	7
180	4.5
210	3
240	3
270	2
300	5
330	14

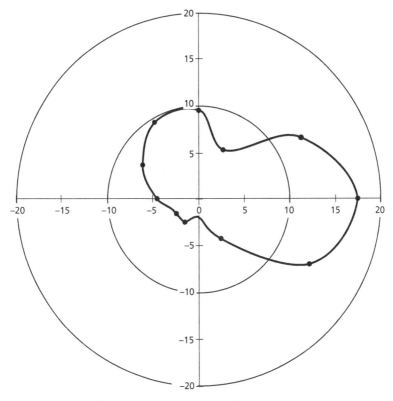

Fig. 6.10 The data from Table 6.3 plotted in polar form.

various locations around the equator (the non-dipole field is that part of the magnetic field which can't be explained as due to a simple bar magnet). Now, the data at a longitude of 330° is only 30° from the data at 0° but, on an x–y plot, it would appear at the opposite end of the graph. Using a polar plot avoids this problem (Fig. 6.10). In the polar plot, the longitude is plotted around the circumference of a circle whilst the field strength is given by the distance from the plot centre (i.e. the stronger the field the further the point is from the centre).

6.5 Equal interval, equal angle and equal area projections of a sphere

This section deals with the problem of plotting data measured on the surface of a sphere onto a flat sheet of paper. This problem occurs, for example, in map making when it is necessary to represent a large portion of the Earth's surface by a map in an atlas. This cannot be done without distortion and there are therefore a large number of different ways of doing this, each of which has advantages and disadvantages. This section will deal with three

Apparent dip	Azimuth
44	11
12	305
31	79
42	2
21	318
34	337
7	112
39	352

Table 6.4 Apparent dip measurements from a single bed at eight different locations. The azimuth gives the directions along which the apparent dips were measured.

very similar methods which are widely used in structural geology, crystallo-graphy, earthquake seismology and many other branches of Earth science. These **projections** are useful whenever information about directions in three dimensions is plotted and they enable many, otherwise complex, manipulations to be carried out relatively simply. There are subtle differences between the methods used in different branches of geology but there is a core of ideas and methods which is common throughout. In this section I introduce some of these ideas but, it must be emphasized, application in particular fields has much more extensive uses than those described here. This section is very much a starting point for the more detailed discussions you will meet in specific geological sub-disciplines.

Consider a bed which outcrops at various locations and whose apparent dip has been measured along a different orientation at each of the different locations. Table 6.4 lists such a series of apparent dip measurements together with the directions in which these dips were measured. Is the bed a simple planar dipping one or is the bed folded in some way? If the bed is planar, what is the true dip and dip direction?

Figure 6.11 illustrates the starting point for resolving these issues. This diagram assumes that the bed is indeed a simple dipping planar bed rep-resented by the dipping plane in this figure. The arrows drawn on the plane represent measurements of apparent dip of this surface taken in various ori-entations. A sphere is drawn with its centre on the plane. The intersection of the plane and sphere is a **great circle**. A great circle is any circle on the surface of a sphere whose centre lies at the centre of the sphere. Thus, the equator and lines of longitude on the Earth's surface are great circles but lines of latitude are not. The individual apparent dip measurements start at the centre of the sphere and intersect the sphere at points on the great circle. If you think of the sphere as having lines of latitude and longitude in a similar fashion to the Earth, each point plots at a 'southerly' latitude equal to the apparent dip and at a longitude equal to the azimuth. Thus, if we plot the projection of our dip measurements onto a sphere and if the resulting points lie along a great

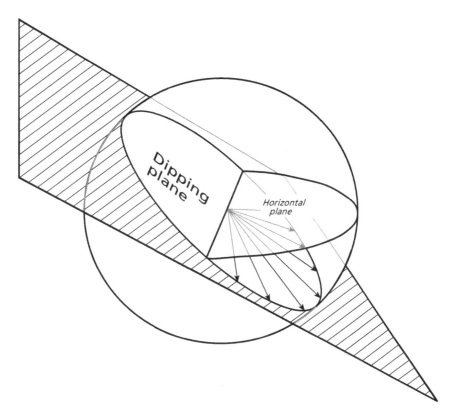

Fig. 6.11 Spherical projection of apparent dip measurements. These will lie along a great circle if the bed is planar.

circle, the measured bed is a simple dipping plane. On the other hand, if this **spherical projection** of the dip measurements is not a great circle, the bed is not planar. Note that, because dips are always measured below the horizontal, only half of all possible orientations are represented. Thus, the spherical projection of the dip data should actually define a semicircle (i.e. the lower half of the great circle).

The problem now is that plotting and performing measurements on a sphere is not very convenient. A solution would be to plot the data using a projection which represents the surface of the sphere on a flat sheet of graph paper.

A method for doing this, bearing in mind that we only have to deal with the lower hemisphere, is to plot onto polar graph paper. To do this, the azimuth is plotted around the circumference of the plot and distance, r, from the centre of the plot is used to represent dip. There are, however, many ways of doing this. The simplest is just to let the distance from the plot edge be proportional to the dip. In other words, dip increases linearly from zero at the plot circumference to 90° at the plot centre. Thus, if a point on the plot

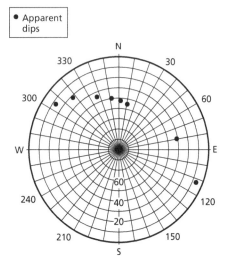

Fig. 6.12 Equal interval polar plot of the data from Table 6.4.

represents a dip of ϕ, the corresponding distance, r, from the plot centre is given by

$$r = R(90 - \phi)/90 \qquad (6.1)$$

where R is the plot radius. For example, a dip of 45° would plot at a distance of

$$r = R(90 - 45)/90$$
$$= R/2$$

i.e. half way between the plot edge and the plot centre. Similarly, a dip of 0° would be at the plot edge (i.e. $r = R$) and a dip of 90° would be at the graph centre (i.e. $r = 0$).

> **Question 6.5** Using Eqn. 6.1, calculate how far apart two points with the same azimuth but dips of 10° and 20° are. Assume the graph radius, R, is 90 mm. Repeat this calculation for dips of 70° and 80°.

Using this method for plotting the measurements in Table 6.4 results in Fig. 6.12 and is known as an **equal interval projection**. The data certainly looks as if it might lie along a semicircle but we have the problem that we don't really know whether this is half of a great circle or just some other fairly smooth curve.

Figure 6.13 shows a **stereographic** or **equal angle projection** of the apparent dip data. This is very similar to the equal interval plot (Fig. 6.12) except that the distance between the circles representing dip is not constant. Note, for example, that the distance between the 0° dip and 10° dip circles is greater

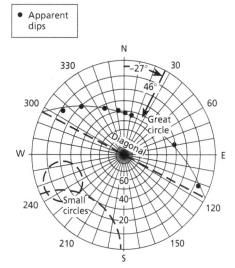

Fig. 6.13 Equal angle (stereographic) projection of the data from Table 6.4. A great circle is also shown which passes close to the apparent dip measurements. Two example small circles are also shown.

than the distance between the 70° dip and 80° dip circles. In this case, the distance from the plot centre is given by

$$r = R \tan[(90 - \phi)/2] \tag{6.2}$$

Thus, a dip of 45° would plot a distance

$$r = R \tan[(90 - 45)/2]$$
$$= R \tan(22.5)$$
$$= 0.414 R$$

which is significantly closer to the centre than half way out (i.e. closer to the centre than in the equal interval plot). Applying Eqn. 6.2 to the cases of zero dip and 90° dip gives a result of $r = R$ and $r = 0$, respectively (the same as for the equal interval plot).

> **Question 6.6** Repeat question 6.5 using Eqn. 6.2. How do these results compare to the equal interval case?

The equal angle projection has two important properties. Firstly, angles measured on the projection are the same as angles on the surface of the sphere. This is particularly useful in crystallography. Secondly, circles drawn on the surface of the sphere project as circles on an equal angle plot and this will be useful for solving our apparent dip problem. Figure 6.13 shows three examples of projections of circles. Two of these are projections of **small circles** (i.e. circles on the surface of a sphere which are not great circles). A great circle is also plotted and is an arc of a circle whose start and end locations define a diagonal to the plot since the two points where it crosses

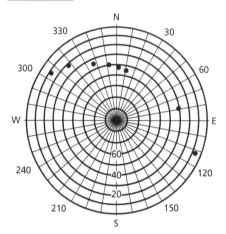

Fig. 6.14 Equal area plot of the data from Table 6.4.

the 0° dip line on the sphere must be opposite each other. In particular, the great circle plotted has been chosen to pass as close as possible through the dip/azimuth data.

This great circle clearly passes quite well through the data points so we can say that the apparent dip measurements are, indeed, taken from a simple planar dipping bed. The true dip and dip direction can also now be found. A dip measurement made in any direction other than the true direction of dip must be smaller than the true dip. Hence, the true dip corresponds to the maximum dip crossed by the great circle. This occurs at the point shown and the true dip and dip direction can be read off as 46° in a direction 27° E of N.

The equal angle projection has the disadvantage that it distorts areas. A figure near the circumference of the plot will plot as an area four times larger than an identical figure at the plot centre. Figure 6.14 shows an **equal area projection** of the apparent dip data. In this case the concentric circles get closer together towards the edge of the plot and r is given by

$$r = \sqrt{2}R \sin[(90 - \phi)/2] \tag{6.3}$$

This time, a dip of 45° will plot at a distance

$$r = \sqrt{2}R \sin[(90 - 45)/2]$$
$$= \sqrt{2}R \sin(22.5)$$
$$= 0.541R$$

which is significantly further from the centre than half way out (i.e. further out than in the equal interval plot). Applying Eqn. 6.3 to the cases of zero dip and 90° dip gives a result of $r = R$ and $r = 0$, respectively (the same as for the equal interval plot).

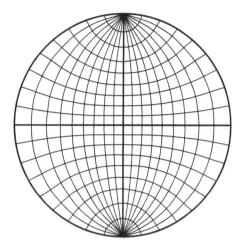

Fig. 6.15 Equatorial net which has great circles and small circles plotted on it.

> **Question 6.7** Repeat question 6.5 using Eqn. 6.3. This time compare your results to both the equal interval and equal angle plots.

The equal area plot has the property that equal areas on the surface of the sphere project as equal areas on the plot. The disadvantage is that circles and angles are now distorted. The equal area projection is frequently used in structural problems rather than the equal angle projection since the density of plotted points is often important and this is distorted by the equal angle projection.

To make it easier to find great circles (and indeed small circles) on these projections, a slightly different type of display from the polar plots is normally used. These are called **equatorial nets** and are drawn with great circles and small circles already plotted upon them. Figure 6.15 shows the equatorial net for the equal angle case. This is known as a **Wulff net**. The equivalent plots for the equal interval and equal area projections are called **Kavraiskii nets** and **Schmidt nets**, respectively, and have a similar appearance to the Wulff net. In Fig. 6.15, the great circles and small circles cut the vertical and horizontal axes at 10° intervals and cut the circumference of the net at 10° azimuth intervals (these lines will usually be drawn at even finer intervals on these nets).

These nets are extremely useful, but plotting a point on an equatorial net is slightly more involved than on the polar type plots. Plotting the position of data points is normally achieved using an equatorial net mounted on a board with a drawing pin through the centre of the net and into a piece of tracing paper placed over the net. This arrangement allows the tracing paper, upon which the data is plotted, to be rotated above the net. Figure 6.16 (a, b and c) shows how to plot one of the data points from Table 6.4 (dip 31°, azimuth 79°) as follows. Figure 6.16a shows a Wulff net with North marked onto the

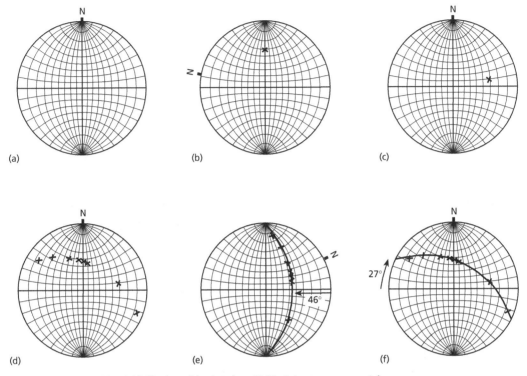

Fig. 6.16 Plotting of the data from Table 6.4 onto an equatorial net.

tracing paper at the top of the net. In Fig. 6.16b the tracing paper is rotated anticlockwise by 79° and the point is plotted 31° down from the top of the net. Rotating the tracing paper so that North is again at the top results in the point being correctly positioned over the net (Fig. 6.16c).

Repeating this procedure for all of the data points from Table 6.4 results in Fig. 6.16d. This figure can then be rotated until the points lie along a great circle (Fig. 6.16e). The distance of this great circle from the plot edge gives the maximum dip value (46°). Finally, rotating the paper again so that North is uppermost results in Fig. 6.16f in which the data and the great circle appear in their correct positions and it can be seen that the dip direction is 27° E of N. Exactly the same set of manipulations can be performed with either the Kavraiskii or the Schmidt nets and, indeed, for many purposes these nets are interchangeable.

The above is only one example of the many problems which these projections can solve. Other examples are: determining the axes of folded structures; determining the earlier orientations of structures which have been multiply deformed; determining detailed earthquake mechanisms; characterizing and identifying crystal structures. This list is very far from being exhaustive. Quite a few of these involve plotting the **poles** of a surface rather than a direction lying on a surface. Poles are outward pointing **normals** to the surface

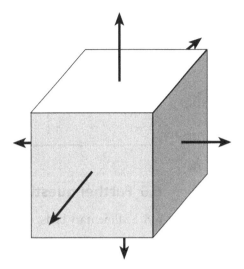

Fig. 6.17 The 6 poles to the faces of a cube.

(i.e. lines at right angles to the surface). In structural geology, downward pointing normals are used instead. Figure 6.17 shows the poles of the surfaces forming a simple cube. As you can see, if this cube has its upper and lower faces horizontal, these poles would plot in a spherical projection with two points at the poles of the sphere and with the remaining four points around the equator.

A small complication with transferring this data to a stereographic (or other) projection is that we have points plotted in both the upper and lower hemispheres whereas a stereographic projection represents only one hemisphere. The solution is to use two projections, one for the upper hemisphere and one for the lower. In practice, both sets of points are plotted on one graph and the difference between them is indicated by using dots for the upper hemisphere points and circles for the lower hemisphere points. Thus, the stereographic projection of the cube poles produces Fig. 6.18.

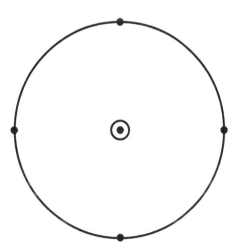

Fig. 6.18 Stereographic projection of the poles from Fig. 6.17. The top and bottom faces plot as the dot and circle, respectively.

Age (My)	Area (10^6 km^2)
> 450	91.1
> 900	50.0
> 1350	35.4
> 1800	26.7
> 2250	7.3
> 2700	1.1

Table 6.5 Area of continental basement which is older than a given age. *Source*: Hurley, P. and Rand, J. (1969). Pre-drift continental nuclei. *Science*, **164**, 1229–42.

6.6 Further questions

6.8 Sedimentary beds, when folded, can have three types of geometry:

(i) Planar (i.e. not folded at all);
(ii) Cylindrical (i.e. folded around one axis, think of a towel hanging over a towel rail);
(iii) Isoclinal (i.e. dome shaped).
In practice, these are the extreme types of fold and real beds are deformed using a combination of these. Thus, for example, a bed might have a very gentle cylindrical fold which can be thought of as a combination of the planar and cylindrical end members. Another bed might be tightly folded around one axis and be gently folded about another axis at right angles (imagine the towel on the towel rail again but this time the rail itself is bent up in the middle). This would be a combination of cylindrical and isoclinal folding. What type of graph would be suitable for illustrating the above concepts?

6.9 Table 6.5 shows the total area of continental crust which is older than a given age. Plot this data in a variety of ways and decide which, you believe, shows the data best.

6.10 Cross-sections through the Earth taken parallel to the equator are, to a good approximation, circular. However, there are small variations due to mountain ranges, ocean basins etc. Imagine such as section taken at a latitude of, say, 30° N. The difference between a circle and the actual section could be tabulated as a function of longitude. What form of plot would be best for displaying such data?

6.11 Consider Fig. 6.19 which shows a planar bed which has been folded about a single axis. Some of the poles to the bedding are also shown and these point in various directions because of the fold.
(i) What form would you expect a spherical projection of the poles to bedding to take?

Fig. 6.19 See question 6.11.

(ii) How does the direction and dip of a pole relate to the direction and amount of bed dip at any particular point? For example, if a bed dips towards 30° E of N what direction does the pole point? If the same bed has a dip of 20° what is the dip of the pole?
(iii) Determine the poles resulting from the following bed dip data and plot them onto an equatorial net

Dip (°)	Dip direction (° E of N)
70	182
65	170
25	131
40	70
36	90
70	35
40	40
37	73
37	146

(Data from McClay, K. (1987) *The Mapping of Geological Structures.* John Wiley, Chichester.)
(iv) Is this information consistent with being taken from a bed folded around one axis?

6.12 Consider the following data:

Location	Dip	Strike	%Sand	%Limestone	%Marl	%TOC	Age (Ma)
1	44	11	10	90	0	0.01	42
2	12	305	30	30	40	14	36
3	7	112	60	20	20	1	38
4	31	79	5	5	90	50	32

Sketch the best types of graphs for showing the following. Ensure the graphs have well-labelled axes and that all appropriate data is marked in approximately the right locations.

(i) The dip and strike relationships.

(ii) The %sand, limestone and marl on a single graph.

(iii) The %TOC as a function of age.

6.13 Use spreadsheet *Triangle.xls* to replot the data from question 6.4.

6.14 Use spreadsheet *Polar.xls* to replot the data from question 6.11 using equal-interval, equal-angle and equal-area projections.

7 Statistics

7.1 Introduction

This chapter is a very brief introduction to the subject of geological statistics. Statistics is probably the most intensively used branch of mathematics in the Earth sciences. For this reason, even an introduction to the subject fills an entire book and there are a large number of such texts. I do not intend, therefore, to cover this topic in the depth it deserves but to give an introduction which I hope will help ease you into the subject and allow you to go on to other texts with some idea of what to expect.

A major problem with statistics is that it is very easy to mislead. A good example comes from the statisticians' favourite subject, coin-tossing. If a coin is tossed six times it is quite likely that there will be three heads. It is very unlikely that six heads will occur. If I then went on to state that it is more probable that the result will be HTHTHT than HHHHHH (where H represents heads and T tails) I would be seriously misleading you. Both of these events are equally unlikely! The reason that the most likely result is three heads and three tails is that there are a large number of ways of doing this (e.g. HHTHTT, HTHTTH and HHHTTT) whereas there is only one way to get six heads (i.e. HHHHHH). Any particular combination of heads and tails is as likely as any other. Other ways in which statistics can mislead are more subtle and even experts can, and do, make very serious errors. However, don't let me put you off statistics. If you work carefully and thoughtfully, statistics can produce results that could not be obtained in any other way.

> **Question 7.1** Write down all possible results of tossing a coin four times. Tabulate the results in terms of the number of different ways of obtaining 0 heads, 1 head, 2 heads, 3 heads or 4 heads. What is the most likely number of heads to get?

7.2 What is a statistic?

I have been writing up to now as if everybody knows exactly what a statistic is. However, even this term is popularly misused. A statement such as 'in 1970 the oil refining capacity of Belgium was 32.6 million tonnes per year' is a fact, not a statistic. So what is a statistic? Let me start with an example of a situation in which statistics might be useful.

Consider a pebbly beach. How would you go about determining the typical composition, mass and length of the pebbles on this particular beach? If I were to pick up one pebble from this beach I would have a **specimen** from the beach. This would probably tell me the composition of some of the pebbles. However, this specimen might be very untypical. A better way to get information would be to pick up one hundred or one thousand pebbles from random locations on the beach. I would then have a **sample** from the beach. This would give me a much better idea of the most common rocks the pebbles were produced from and their typical masses and lengths. Finally, I could examine (in principle) all the pebbles from the beach. This is the **population** of all pebbles from this beach and I could then make definitive statements about the composition of the beach. To recap, a specimen is one object, a sample is a number of objects and a population is all the relevant objects. Note that the word sample is frequently used in geology to denote a specimen (e.g. 'a sample of sandstone' meaning a single piece of sandstone). This is confusing and I recommend that you use the word 'specimen' whenever possible.

Question 7.2 If I have six books from a library containing 10 000 books, do these six form a specimen from the library, a sample from the library or the library population?

Now we can return to the idea of a statistic. Is the average mass of a pebble a statistic? This depends on whether this average is determined from a sample of pebbles or from the total population of pebbles. The average of the population is a **parameter** of the beach and is a simple fact (just like the Belgian oil refining capacity). The average of a sample, on the other hand, is a **statistic**; it is an attempt to estimate the average mass of all the pebbles by calculating the average mass of some of the pebbles. In other words, a statistic is an estimate of a parameter based upon a sample of the population. As another example, consider voting patterns in an election. The estimates of voting intentions obtained by polling organizations before the election itself are statistics (they are based on questioning a small minority of voters), whereas the final official result is a parameter of the election.

Returning to the beach example, the way in which the masses vary from pebble to pebble is described by many parameters in addition to the average mass. For example, the pebble masses may all be very close to one another or they may be widely different. One parameter which quantifies this is called the **standard deviation**. This will be defined more precisely in the next section. Another parameter is called the **skew** of the population and this tells us whether there are more pebbles which are heavier than the average or more pebbles which are lighter. All of these parameters would normally be estimated from a sample of the population. Each of the resulting estimates of a parameter is a statistic.

Whether or not these statistics are good estimates depends on how well the sampling was performed and also on the size of the sample. Two pebbles picked up from one place on the beach are unlikely to yield a good estimate of the average mass. One hundred pebbles picked up at random from all over the beach will give a much better estimate. Designing an experiment or fieldwork so that the information collected represents a good sample is an important part of a scientific approach to a problem.

7.3 Commonly encountered parameters and statistics

Table 7.1 shows the masses of 100 pebbles from a beach. From this sample we might wish to get an idea of the typical mass and also the spread of masses. Each of these can be quantified in several different ways.

Table 7.1 The masses of 100 pebbles sampled from a beach.

Pebble number	Mass (g)	Pebble number	Mass (g)	Pebble number	Mass (g)	Pebble number	Mass (g)
1	374	26	294	51	284	76	287
2	389	27	256	52	403	77	340
3	358	28	359	53	341	78	401
4	395	29	352	54	435	79	422
5	371	30	330	55	307	80	369
6	334	31	269	56	420	81	379
7	224	32	355	57	342	82	432
8	335	33	283	58	331	83	368
9	256	34	301	59	331	84	338
10	340	35	346	60	331	85	327
11	374	36	393	61	290	86	433
12	423	37	386	62	383	87	370
13	338	38	338	63	370	88	343
14	373	39	380	64	302	89	450
15	342	40	357	65	394	90	318
16	242	41	326	66	329	91	384
17	318	42	403	67	324	92	355
18	454	43	317	68	283	93	366
19	346	44	301	69	355	94	324
20	408	45	394	70	311	95	353
21	403	46	407	71	265	96	277
22	384	47	350	72	364	97	359
23	397	48	375	73	322	98	400
24	307	49	303	74	283	99	314
25	409	50	384	75	367	100	389

The typical mass can be described by the **mean**, $\bar{w}$,

$\bar{w}$ = (Total mass of the sample)/(number of pebbles)
    = 35 018/100 = 350.18 g   (7.1)

for the pebbles in Table 7.1. Frequently, this computation will be described using the following notation:

$$\bar{w} = \frac{1}{N}\left(\sum_{i=1}^{N} w_i\right) \qquad (7.2)$$

where w_i is the mass of the ith pebble and N is the number of pebbles in the sample (i.e. w_1 is the first mass in Table 7.1 (374 g), w_2 is the 2nd mass (389 g) and so on). The symbol 'Σ' (sigma) is an instruction to add together the w_i's (i.e. add the masses of the pebbles together). The $i = 1$ below the Σ, and the N above, indicate that all items numbered between 1 and N have to be added (i.e. the mass of 100 pebbles in our example). Finally, the result of this addition is divided by N (i.e. 100 in our case). Equation 7.2 may be rewritten in words as 'the average mass may be found by summing the masses of N pebbles and dividing by N'. This notation will be used repeatedly throughout this chapter so make sure that you understand it. Note also that drawing a line over a letter (e.g. $\bar{w}$), to denote an average, is a very common convention.

An alternative way of quantifying the typical mass is to use the **median** value. This is obtained by **ranking** the pebbles from the heaviest to the lightest and taking the central value. If we had 5 pebbles, the central one would be the third heaviest pebble (also the third lightest). Similarly, for 99 pebbles the median would be the mass of the fiftieth heaviest pebble. However, for an even number of pebbles there is no central pebble. For example, for 4 pebbles the second and third are equally close to being central. In such cases the procedure is to average the two most central values. For 100 pebbles we must average the mass of the fiftieth and fifty-first pebble. Ranking of the beach sample in Table 7.1 results in Table 7.2. Pebbles 50 and 51 are then 353 g and 352 g, respectively. Thus, the median mass is the average of 353 and 352, i.e. 352.5 g. Note that this is different to the average mass.

So much for statistics which indicate the typical mass. What about other aspects of the distribution of pebble masses? For example, the pebbles might all have very similar masses or the masses might be very widely dispersed. What is needed is a measure of **dispersion**. A very simple way to indicate this would be to give the **range** of values, i.e. the lowest and highest masses in the sample. In the case of Table 7.1 (or better, Table 7.2) the range is from 224 g to 454 g. However, the heaviest and lightest pebbles might be very untypical. It would be better to use a measure of the spread of masses which is

Table 7.2 Pebbles ranked according to decreasing mass.

Rank	Mass (g)	Rank	Mass (g)	Rank	Mass (g)	Rank	Mass (g)
1	454	26	384	51	352	76	318
2	450	27	384	52	350	77	318
3	435	28	383	53	346	78	317
4	433	29	380	54	346	79	314
5	432	30	379	55	343	80	311
6	423	31	375	56	342	81	307
7	422	32	374	57	342	82	307
8	420	33	374	58	341	83	303
9	409	34	373	59	340	84	302
10	408	35	371	60	340	85	301
11	407	36	370	61	338	86	301
12	403	37	370	62	338	87	294
13	403	38	369	63	338	88	290
14	403	39	368	64	335	89	287
15	401	40	367	65	334	90	284
16	400	41	366	66	331	91	283
17	397	42	364	67	331	92	283
18	395	43	359	68	331	93	283
19	394	44	359	69	330	94	277
20	394	45	358	70	329	95	269
21	393	46	357	71	327	96	265
22	389	47	355	72	326	97	256
23	389	48	355	73	324	98	256
24	386	49	355	74	324	99	242
25	384	50	353	75	322	100	224

determined by all of the pebbles in the sample rather than a small minority. One such measure is the **mean square deviation from the mean**. This is also called the **variance**. For the total population of pebbles this is denoted by σ^2 and is defined as:

$$\sigma^2 = \overline{(\text{mass} - \text{average mass})^2} \tag{7.3}$$

where the bar over the expression indicates that the average value of this quantity should be calculated. In other words, we first find the average pebble mass and then calculate the difference between this and the mass of each individual pebble. The result is then squared which gives the square deviation from the mean. Finally, the average value of this for all the pebbles is found. The deviation of each pebble from the mean is squared since some of the deviations are negative (mass less than average) and some are positive (mass higher than average), leading to an average deviation of zero. Squaring all the deviations from the mean ensures that the average of a series of positive numbers is found which will, of course, also be a positive number. Notice that if

the masses are all very similar then they will all be very close to the mean leading to a small value for the variance. If, on the other hand, the masses differ
widely from one another then some of these masses will be a long way from
the average value and the variance will be much larger. The standard deviation, σ, which is simply the square root of Eqn. 7.3, could also be used to
indicate the range of values in the population.

However, it would be better to have a measure of distribution width
based upon a sample rather than the entire population. An obvious candidate
would be the **sample variance**, s^2, i.e. apply Eqn. 7.3 to a sample rather than
the population. In terms of the notation introduced earlier this gives

$$s^2 = \frac{1}{N}\left(\sum_{i=1}^{N}(w_i - \bar{w})^2\right) \tag{7.4}$$

where $\bar{w}$ is the average mass defined by Eqn. 7.2. There is a slightly easier
method for calculating s^2 since Eqn. 7.4 can be rearranged to give

$$s^2 = \overline{w^2} - (\bar{w})^2 \tag{7.5}$$

i.e. the mean of the squared masses minus the square of the mean mass
(proof of this is given as an exercise at the end of the chapter). Using the
figures from Table 7.1, the square of the pebble 1 mass is $374 \times 374 =$
$139\ 876\ g^2$. Repeating this for all pebbles and taking the average then gives
a mean square mass of $124\ 876\ g^2$. The square of the mean mass, on the
other hand, is $350.18^2 = 122\ 626\ g^2$. Thus, using Eqn. 7.5, the sample variance is $124\ 876 - 122\ 626 = 2250\ g^2$.

As a measure of the width of the distribution this number has several disadvantages. The first problem is that $\bar{w}$ itself has been estimated from the
sample. In fact, the estimate of $\bar{w}$ obtained from Eqn. 7.2 has the property
that it is the value which minimizes the sample variance. If another value is
used in Eqn. 7.4, in place of $\bar{w}$, a larger value for s^2 is always obtained. Hence,
if the true population mean were substituted into Eqn. 7.4 instead of its
estimate, $\bar{w}$, a larger value for s^2 would result. Equation 7.4 is therefore
biased towards a smaller value than the true one. To counteract this effect the
unbiased estimate, $\hat{s}^2$ is used instead, where

$$\hat{s}^2 = [N/(N-1)]s^2 \tag{7.6}$$

Note that, for large sample sizes, this increases the variance very slightly,
whereas for smaller sample sizes this variance estimate is significantly larger
than s^2. A formal proof that $\hat{s}^2$ is a better estimate of the population variance
than s^2 is beyond the scope of this chapter, but you should be able to see that
it has the effect of altering s^2 in the right direction (i.e. it increases it by an
amount which depends upon the sample size).

Table 7.1 has a value for N of 100, the figure calculated above for s^2 then produces an estimate of the population variance of $\hat{s}^2 = 100 \times 2250/99 = 2273 \text{ g}^2$.

Another problem with using variance to measure distribution width is that the final number (in this case 2273 g²) is not very understandable. What does this result actually mean? Perhaps the simplest way to look at this is to use the variance for comparison purposes. Take two samples of 100 pebbles from two different beaches. If the first beach has a larger variance for the masses than the second, the pebbles on the second beach tend to have more similar masses to each other than those from the first beach.

The 'interpretability' of variance is further improved by taking its square root. This gives an estimate, $\hat{s}$, of the population standard deviation σ. In the case of our data this yields an answer of $\sqrt{2273} = 47.7$ g. For reasons that are covered in more detail later (see Section 7.5), this result implies that around 68% of all pebble weights should fall within 47.7 g of the mean value (350.2 g). Thus, 68% of all weights should fall between 302.5 g and 397.9 g. In fact, out of the 100 measurements in Table 7.1, 65 fall in this range which is, of course, 65% of the total. Thus, the theoretical prediction that 68% of the pebble weights should fall in this range is not at all bad. Standard deviation is therefore a very simple way of describing the range of values in your data. A large standard deviation implies a wide spread of values whilst a small standard deviation implies a small spread.

Question 7.3 Using the first 10 values from Table 7.1, calculate:
(i) the sample mean;
(ii) the sample median;
(iii) the sample variance.
Also estimate:
(iv) the population variance;
(v) the standard deviation.
Compare these results to those obtained above from the sample of 100 pebbles.

There are many other parameters and statistics which could be calculated for given populations and samples. However, the most important are undoubtedly the mean and the standard deviation and these are the ones with which you should be most familiar.

7.4 Histograms

It is useful to have a method for displaying, for example, the distribution of pebble masses in Table 7.1 graphically. The simplest such method uses the

Range (g)	Number
201–250	2
251–300	12
301–350	35
351–400	36
401–450	14
451–500	1

Table 7.3 Number of pebbles in Table 7.1 which fall into 50-g-wide classes.

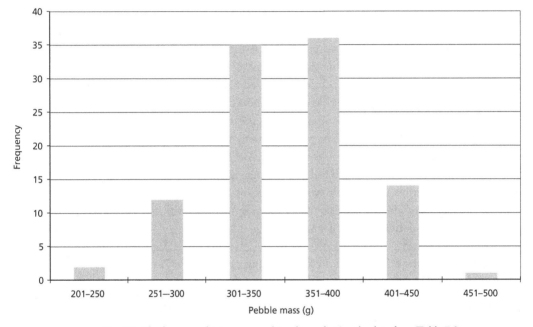

Fig. 7.1 The frequency histogram resulting from plotting the data from Table 7.3.

frequency histogram which allows the general properties of the distribution to be visualized. To construct such a plot we must first count the number of occurrences of pebble masses within specified ranges. For example, there are 2 pebbles in Table 7.2 with masses between 201 g and 250 g. Table 7.3 lists the number of pebbles with masses in 50-g-wide **classes** from 201 g to 500 g. Such a table is called a **frequency distribution**. If these figures are plotted as a bar chart, the result is a frequency histogram (Fig. 7.1). From this we can instantly see that the masses most commonly fall between 300 g and 400 g.

It was pointed out in Chapter 6 that, if your data is a function of a cyclic variable such as direction or longitude, it is best represented by a polar plot. The same is true for histograms. For example, cross bedding and ripple marks in sandstones can be used to indicate the **palaeocurrent direction**, i.e. the direction of transport of ancient rivers or submarine currents. Such data will

Table 7.4 Number of palaeocurrent
measurements as a function of direction.

Direction range (° E of N)	Number of measurements
1–30	43
31–60	23
61–90	10
91–120	11
121–150	14
151–180	20
181–210	10
211–240	4
241–270	15
271–300	20
301–330	40
331–360	36

usually have considerable scatter due to uncertainties in measurement and the effect of local topographic features. Thus, if the general direction of transport is required, the best course of action is to collect a large number of measurements and then to plot these on a histogram. Table 7.4 lists the frequency data from such a series of measurements.

Now, an obvious way to plot this data is as a histogram on polar graph paper. In other words, plot the frequency as a function of direction such that direction is represented by angle around the plot and the frequency is proportional to distance from the plot centre. This yields a **rose diagram** (Fig. 7.2) from which it is very clear that the main current direction was roughly NNW. Spreadsheet *Rose.xls* can be used to plot this, and other similar, data.

7.5 Probability

Probability is a central concept in statistics. In essence, the idea is very simple. If I perform a very large number of measurements on field data or experimental data then I can determine how often a particular result is obtained. This will then allow me to predict the probability that a particular result will occur in any future measurement. Thus, if I toss a dice 1000 times and the number two occurs 400 times, I can predict that the probability is 0.4 of two being the result of my next throw of the dice. I can also conclude that the dice is probably loaded. Note that an event which has a probability of one is certain to occur whilst an event whose probability is zero will never occur.

Similarly, for the data shown in Table 7.3 and Fig. 7.1, the most probable weight range (351–400 g) occurs in 36 cases out of 100, i.e. 36% of the time. In other words, an estimate of the probability of a pebble, picked up at random from the beach, being in the range 351–400 g is 0.36. Repeating this

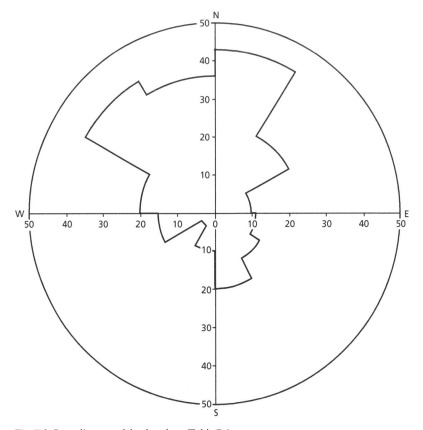

Fig. 7.2 Rose diagram of the data from Table 7.4.

Range (g)	Probability
201–250	0.02
251–300	0.12
301–350	0.35
351–400	0.36
401–450	0.14
451–500	0.01

Table 7.5 Estimates of probability for each pebble mass range using the data from Table 7.1. Note that results are simply those from Table 7.3 divided by the number of specimens (100).

procedure for the entire table leads to Table 7.5 which is a **probability distribution**. The results can then be plotted in a new histogram (Fig. 7.3). Note that the shape of this is identical to Fig. 7.1 except that the vertical scale has been shrunk by a factor of 100 (i.e. divide by the size of the sample).

This probability distribution can be compared to various theoretical distributions. The most important of these is the **normal distribution** otherwise known as the **Gaussian distribution**. This has the form

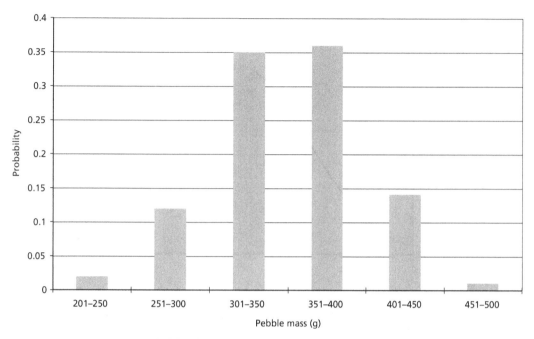

Fig. 7.3 Probability distribution histogram for the data from Table 7.5.

$$P(x) = \frac{\exp[-(x - \bar{x})^2/2\sigma^2]}{\sqrt{2\pi\sigma^2}} \quad\quad (7.7)$$

where $P(x)$ is called the *relative probability* of obtaining the value x,
$\bar{x}$ is the average of all x values,
σ is the standard deviation of the distribution.
A graph of this function is shown in Fig. 7.4 for the case of mean equal to 5.0 and standard deviation equal to 2.0. Note that the maximum probability occurs at the mean value, that the curve is symmetrical about the mean and that a large fraction of the area under the graph occurs between $\bar{x} - \sigma$ and $\bar{x} + \sigma$ (i.e. the dark grey area).

The probability of obtaining values within a specified range is governed by two things. Firstly, the higher the graph is within that range, the higher the probability is. Thus, given the graph shown in Fig. 7.4, you are more likely to obtain a value between, say, 5 and 6 (where the graph is high) than you are between 1 and 2. Secondly, the probability of obtaining a value within a specified range increases as the width of the range increases. Thus, you are more likely to obtain a value between 5 and 7 than between 5 and 6 because more possibilities are included in the second case. In fact, the relative probability distribution is defined such that the probability of obtaining a value within a given range is given by the area under the graph over that range. For example, the probability of obtaining a value between 1.0 and 2.0 is

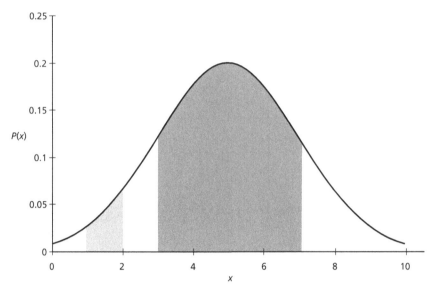

Fig. 7.4 Gaussian probability distribution for a mean of 5.0 and standard deviation of 2.0. The area of the light grey shaded region gives the probability of a specimen lying in the range 1–2. Similarly, the dark grey area gives the probability of a specimen lying between 3–7. The total area under the graph is 1.0.

given by the light grey area shown in Fig. 7.4 whilst the (much greater) probability of obtaining a value between 3 and 7 is given by the dark grey area. Given this definition, the total area under the graph is 1.0 since the probability of obtaining some value is 1.0. This is, indeed, the case for the Gaussian distribution defined in Eqn. 7.7.

So, how can such areas be found? The simplest method is to use a table showing area under the curve as a function of multiples of standard deviation from the mean (e.g. Table 7.6). From such a table it can be seen that the area of the dark grey region in Fig. 7.4 is 0.683. Similarly, the area under the curve within two standard deviations (i.e. between 1.0 and 9.0 in the example shown in Fig. 7.4) is 0.954.

The table can also be used to find areas such as the grey zone in Fig. 7.4. To do this, you should note that 1.0 is two standard deviations from the mean whilst 2.0 is 1.5 standard deviations from the mean. From Table 7.6 it can be seen that the area within 1.5 σ is 0.866 whilst the area within 2 σ is 0.954. Thus, the area between 1.5 σ and 2.0 σ is 0.954 − 0.866 = 0.088. However, there are two such zones, one between 1.0 and 2.0 and the other between 8.0 and 9.0 (see Fig. 7.5). The area between 1.5 σ and 2.0 σ is shared equally between these two regions and thus the area of the zone between 1.0 and 2.0 is half of 0.088 (i.e. 0.044). Thus, if a process has a probability distribution like that shown in Fig. 7.4, the probability of obtaining a result between 1.0 and 2.0 is 0.044.

Table 7.6 Area under the Gaussian curve as a function of number of standard deviations from the mean. For example the probability of lying within 2 sd of the mean is 0.954.

Number of standard deviations	Area	Number of standard deviations	Area	Number of standard deviations	Area
0.00	0.000				
0.10	0.080	1.10	0.729	2.10	0.964
0.20	0.159	1.20	0.770	2.20	0.972
0.30	0.236	1.30	0.806	2.30	0.979
0.40	0.311	1.40	0.838	2.40	0.984
0.50	0.383	1.50	0.866	2.50	0.988
0.60	0.451	1.60	0.890	2.60	0.991
0.70	0.516	1.70	0.911	2.70	0.993
0.80	0.576	1.80	0.928	2.80	0.995
0.90	0.632	1.90	0.943	2.90	0.996
1.00	0.683	2.00	0.954	3.00	0.997

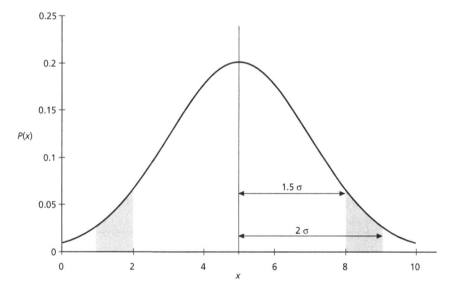

Fig. 7.5 The area under the Gaussian curve between 1.5 and 2.0 standard deviations from the mean. Note that there are two such zones.

Thus, if a population variable has a Gaussian probability distribution, the probability of obtaining results within specified ranges can be calculated. However, the Gaussian distribution function is an idealized model. Real populations are often approximately Gaussian but never exactly so. How good is the Gaussian model at predicting the probabilities shown in Table 7.5 for the pebble weights on our beach? Now, the mean and standard deviation

Table 7.7 Comparison of the probabilities estimated in Table 7.5 with a Gaussian distribution having the same mean and standard deviation.

Range (g)	Measured probability	Range (multiples of σ)	Gaussian probability
201–250	0.02	−3.10 to −2.06	0.019
251–300	0.12	−2.06 to −1.02	0.134
301–350	0.35	−1.02 to 0.02	0.354
351–400	0.36	0.02 to 1.06	0.347
401–450	0.14	1.06 to 2.10	0.127
451–500	0.01	2.10 to 3.13	0.017

for the pebble weights have been estimated to be about 350 g and 48 g, respectively. Using these values, Table 7.6 can be used to predict the probabilities in various ranges of weights. The results are shown in Table 7.7 together with the measured probabilities shown in Table 7.5 (in fact I have used a slightly more detailed table than Table 7.6). For this data set the Gaussian model seems to be pretty good.

> **Question 7.4** The range from 401 g to 450 g starts 1.06 standard deviations above the mean and ends 2.10 standard deviations above the mean. Using Table 7.6 and these values, estimate the Gaussian probability of a pebble weight lying in this range. N.B. To get the area under the curve within 1.06 standard deviations (call it $P_{1.06}$) assume that it is given by the expression
>
> $$P_{1.06} \approx 0.6\, P_{1.1} + 0.4\, P_{1.0}$$
>
> i.e. an average of the probabilities corresponding to $1.0\,\sigma$ and $1.1\,\sigma$ weighted towards the $1.1\,\sigma$ probability. Check the result using spreadsheet *Gauss.xls*.

7.6 Best fit straight lines

So much for statistical analysis of a single variable. What about analysis of two related variables? As discussed in Chapter 2, it is very common for graphs of the relationship between pairs of geological variables to be well approximated by straight lines. However, the fit is never perfect. Thus, a straight line must be found which passes as close to all the data points as possible. The problem of how to find this line is ideally suited to a statistical treatment.

First, we have to define what we mean by a **best fit** straight line. The usual definition is that the mean square difference between the data and the straight

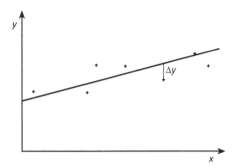

Fig. 7.6 A straight line drawn through some x–y data such that it passes close to all of the points. The deviation, Δy, of one of the points from the line is also shown.

line should be a minimum. Figure 7.6 illustrates this idea. The graph of y as a function of x consists of seven points and a straight line has been drawn which passes close to all of these. The deviation, Δy, of one point from the straight line is also shown. The mean square deviation is found by calculating the square of this distance for all of the points and then finding the average. Now, if the line is a poor fit to the data, Δy for many of the points will be large and the average squared value will also be large. A good fit for the straight line will result in a much smaller average. The best fit straight line is defined as that line which results in the smallest possible **mean square deviation**. The process of finding such a line is called **linear regression**.

We now need a method for estimating the gradient, m, and intercept, c, of this straight line. A formal proof is, again, beyond the scope of this chapter but the result is that the best gradient is given by

$$m = \frac{n \sum xy - \sum x \sum y}{n \sum x^2 - (\sum x)^2} \tag{7.8}$$

In this expression all summations are evaluated for n measurement pairs (x,y). The first summation, $\sum xy$, is simply the sum of all products $x_1 y_1, x_2 y_2$ up to $x_n y_n$. Similarly, $\sum x$ is the sum of all the x values, $\sum y$ is the sum of all the y values and $\sum x^2$ is the sum of all the x values squared. The best estimate for the intercept then follows directly from the fact that the best fit line passes through the point $(\bar{x}, \bar{y})$, i.e. the point defined by the average x value and the average y value. Thus,

$$\bar{y} = m\bar{x} + c$$

giving

$$c = \bar{y} - m\bar{x} \tag{7.9}$$

Table 7.8 shows the age versus depth data used in question 2.11 of Chapter 2 but with the first point excluded. Now, given that the remaining data when plotted seemed to fit a reasonably good straight line, what is the best fit line

Depth (cm)	Age (years)
407	10 510
545	11 160
825	11 730
1158	12 410
1454	12 585
2060	13 445
2263	14 685

Table 7.8 Age versus depth data from question 2.11 but with first data point excluded.

through these points? First, we should construct all the summations given in Eqn. 7.8. In this example, we need to plot age as a function of depth and therefore *Depth* replaces x and *Age* replaces y. The summations required by Eqn. 7.8 are then

$$\Sigma\ depth = 407 + 545 + 825 + 1158 + 1454 + 2060 + 2263$$
$$= 8712 \text{ cm} \tag{7.10}$$

$$\Sigma\ age = 10\ 510 + 11\ 160 + 11\ 730 + 12\ 410 + 12\ 585 + 13\ 445 + 14\ 685$$
$$= 86\ 525 \text{ years} \tag{7.11}$$

$$\Sigma\ depth.age = (407 \times 10\ 510) + (545 \times 11\ 160) + (825 \times 11\ 730)$$
$$+ (1158 \times 12\ 410) + (1454 \times 12\ 585) + (2060 \times 13\ 445)$$
$$+ (2263 \times 14\ 685)$$
$$= 4\ 277\ 570 + 6\ 082\ 200 + 9\ 677\ 250 + 14\ 370\ 780$$
$$+ 18\ 298\ 590 + 27\ 696\ 700 + 33\ 232\ 155$$
$$= 113\ 635\ 245 \text{ cm years} \tag{7.12}$$

$$\Sigma\ (depth^2) = 407^2 + 545^2 + 825^2 + 1158^2 + 1454^2 + 2060^2 + 2263^2$$
$$= 13\ 963\ 148 \text{ cm}^2 \tag{7.13}$$

From Eqns. 7.10 and 7.11, together with the fact that there are seven measurements, the mean depth and mean age are

$$\overline{depth} = 8712/7 = 1245 \text{ cm} \tag{7.14}$$

and

$$\overline{age} = 86\ 525/7 = 12\ 361 \text{ years} \tag{7.15}$$

Thus, substituting results 7.10 to 7.13 into Eqn. 7.8 gives

$$m = \frac{7 \times 113\ 635\ 245 - 8712 \times 86\ 525}{7 \times 13\ 963\ 148 - 8712 \times 8712}$$
$$= 1.91 \text{ years cm}^{-1} \tag{7.16}$$

Substituting this into Eqn. 7.9 gives the best estimate of the intercept as

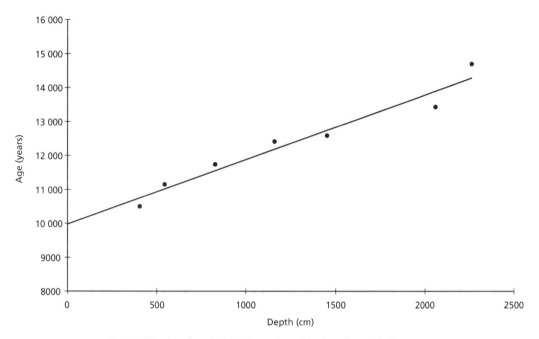

Fig. 7.7 The data from Table 7.8 together with a best fit straight line.

$$c = \overline{age} - (m.\overline{depth})$$
$$= 12\ 361 - (1.91 \times 1245)$$
$$= 9983 \text{ years} \hspace{4cm} (7.17)$$

The data from Table 7.8 is plotted in Fig. 7.7 together with a straight line of gradient 1.91 yr cm^{-1} and intercept 9983 years. As you can see, the fit is remarkably good. Note that, before starting this exercise, I deliberately excluded a point that did not fit the general linear trend. This means that the resultant line is not appropriate at ages close to this point (i.e. for sediments less than about 10 000 years old). However, it was important to exclude this point since fitting a straight line through points which do not have a simple linear trend would be a meaningless exercise.

> **Question 7.5** Calculate the best fit straight line through the data in Table 2.2 of Chapter 2.

The principles behind linear regression can be applied to other functions. For example, spreadsheet *Bfit.xls* is set up to calculate best fit polynomials through datasets and can be used for fitting curves of the form of Eqn. 2.9 (see Section 2.4) for $n = 0, 1, 2, 3$ and 4. Note that for $n = 0$ it will simply calculate the mean of the data, for $n = 1$ it will find the best fit straight line and for $n = 2$ it will fit a quadratic function.

7.7 The importance of error estimates

All data values are wrong! It is impossible to measure anything with infinite precision and so any measured quantity must differ slightly from its true value. A pebble mass may be measured to the nearest gram but is unlikely to really weigh an exact number of grams. Pebble 1 in Table 7.1 is quoted as weighing 374 g but it might be 374.126 547 g or 373.556 321 678 g or any other mass between 373.5 g and 374.5 g.

To make matters worse, many measurements are hard to make even as accurately as the measuring instrument can theoretically manage. Dip and strike measurements, for example, could in theory be measured with a standard geologist's compass clinometer to the nearest degree. In practice it is hard to make measurements this accurately and most field measurements are probably in error by several degrees.

An additional problem is that data values themselves may be inherently variable. The dip of a planar bed will change slightly between nearby locations because no real sedimentary surface is perfectly flat. Hence, any single measurement on that bed is likely to give a dip that differs at least a little from the average bed dip.

Table 7.9 shows dip and strike measurements taken by 20 different students at two different field locations. You can see that, for either of the two locations, the individual dip values vary by as much as 10° whilst the strike values are even more variable. These variations occur for the reasons given above, namely that the measurements are hard to do accurately and, in any case, the dip and strike depend upon the exact spot at which the measurements were taken.

Data without error estimates are useless. Take student number 1's measurements in Table 7.9. An interesting question for the student to ask might be 'is the dip higher at location B than at location A?' From his dips of 22° at location A and 25° at location B it is impossible to say since the increase may be real or may be entirely due to measurement error. Dips differing by 60° are clearly different but what if they differ by 1°? Most geologists would agree that a 1° difference is not significant, but why would they say this? The reason is that 60° is much bigger than any likely uncertainty in the measurement whilst a 1° difference is a lot less than variations caused by measurement error or unevenness of the bed.

Carefully taken dip measurements should be accurate to about 3° (I'll explain how to estimate this in Section 7.8). Hence, with student number 1's data, it is very hard to be sure if the dip has increased between locations. The difference between the dips is uncomfortably close to the uncertainty in the measurements.

Table 7.9 Dip and strike measurements from two locations measured by 20 different students.

Student	Location A		Location B	
	Strike	Dip	Strike	Dip
1	51	22	60	25
2	45	23	54	22
3	44	21	69	22
4	61	24	59	19
5	37	21	58	27
6	11	23	59	16
7	26	23	66	18
8	54	21	62	22
9	42	22	48	23
10	29	22	62	16
11	48	22	53	16
12	34	18	72	19
13	50	26	62	21
14	61	24	69	21
15	36	26	70	26
16	47	30	41	20
17	30	21	59	16
18	54	25	76	22
19	48	21	54	22
20	55	21	64	12
Average	43.15	22.80	60.85	20.25
Standard deviation	12.70	2.57	8.41	3.80
Standard error	2.84	0.57	1.88	0.85

Despite these types of difficulty, most geological data values are given without error estimates. This might be excusable for dip and strike measurements since most geologists have a pretty good (perhaps subconscious) idea of how accurate such measurements are. However, it is good practice to give error estimates whenever possible. Don't trust anyone's numbers if the person responsible cannot give you at least a rough idea of how accurate they might be.

Question 7.6 Cores taken from two different wells, but from the same sedimentary bed, have sand/shale ratios of 0.50 and 0.51, respectively. If it is hard to determine this ratio with an accuracy better than about 0.02, do you think it is safe to infer that the sand/shale ratio is higher in the second well?

7.8 Quantitative estimates of error

Returning to the data in Table 7.9, the fact that there are 20 independent estimates of the bed strike and dip allows much more to be said about the problem of whether the bed has changed orientation between the two locations. To start with, the methods given in Section 7.3 allow average strike/dip and associated standard deviations to be estimated. The results are given in Table 7.9.

As should be clear from Section 7.3, the standard deviation is an estimate of the typical deviation of any given measurement from the mean value. Hence, each individual student's strike measurements have an error of around 10° (strike standard deviations are 12.7° and 8.4° at A and B, respectively) and a dip error of about 3° (dip standard deviations are 2.6° and 3.8° at A and B, respectively). Note that student 1's estimates of the dips at locations A and B (22° and 25°, respectively) differ from each other by an amount which is similar to the measurement uncertainty and so, from student 1's measurements alone, it is impossible to say which dip is larger.

However, the average of the 20 measurements should be much more accurate than the individual values and, in general, should improve as the number of measurements increases. Note that, unlike the impression given by student 1's measurements, the average dip at B is actually less than at A. An estimate of the typical deviation of the sample mean from the true mean is given by the standard error, s_e, where

$$s_e = \hat{s}/\sqrt{N} \tag{7.18}$$

where $\hat{s}$ is the standard deviation and N the number of observations. The resulting s_e estimates are also given in Table 7.9.

The exact meaning of the standard error depends upon the sample size. This is because estimates of a population mean have a **t-distribution**. For large sample sizes the t-distribution is similar to the Gaussian distribution introduced earlier but for smaller sample sizes it is different. Figure 7.8 shows the multiple of the standard error within which there is a 95% probability of the true value lying. This depends upon the **degrees of freedom** (explained below) which, for the problem considered in this section, is given simply by $N - 1$. Hence for our sample size of $N = 20$ there are 19 degrees of freedom implying that the true mean has a 95% chance of lying within $2.1s_e$ of our mean estimate. At location A, for example, the true strike has a 95% chance of lying within $2.1 \times 2.9 = 6.1$ degrees of 43.2°. Similarly, the 95% confidence interval for the strike at B is $60.9 \pm 4.0°$. Hence, this analysis implies that the strike values at the two locations are clearly different.

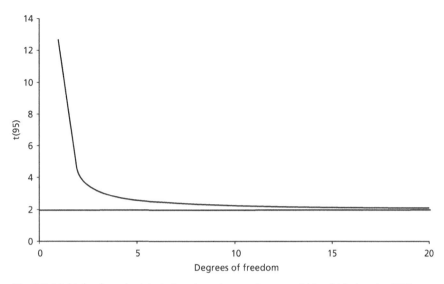

Fig. 7.8 Multiple of standard deviations from the sample mean within which there is a 95% chance of the true mean lying.

> **Question 7.7** Using Fig. 7.8, find the 95% confidence limits for the dips at locations A and B. Check your answer using spreadsheet *Sterr.xls*.

You should have found, from your answer to question 7.7, that the possible range of dips at locations A and B just overlap implying that, even after averaging 20 measurements, we still cannot say which location has the higher dip. A more sophisticated statistical approach (a ***t*-test**, not covered here) would just allow these particular two values to be distinguished. However, the simple approach used here, i.e. comparing 95% confidence ranges to determine whether two values are significantly different, will suffice for many purposes. Note that it is inherently conservative in the sense that only very well separated values will be distinguished as clearly different.

For completeness, a few words need to be said concerning the concept of degrees of freedom. Generally, if there are N specimens in a sample then there is freedom to change N numbers (i.e. there are N degrees of freedom for N measurements). However, in the problems discussed above, we first used the N measurements to determine a mean value. Once a mean has been set it is only possible to change $N - 1$ values since the Nth value is forced to be a particular number if the mean is to be retained. Hence, after estimating both a mean and a standard error, the confidence intervals are determined from the t-distribution for $N - 1$ degrees of freedom. Don't worry too much about this rather abstract point. In practice, statistics books give quite explicit instructions on how many degrees of freedom are appropriate for a particular problem.

Table 7.10 Dip directions for one bed at 20 different locations.

Location	Direction	Location	Direction
1	27	11	14
2	63	12	355
3	10	13	300
4	87	14	96
5	103	15	276
6	256	16	190
7	200	17	191
8	23	18	23
9	17	19	10
10	25	20	5

7.9 Further questions

7.8 Measured dip directions for a particular bed at 20 different locations in a field area are given in Table 7.10.
(i) Use this data to calculate the frequency distribution and estimate the probability distribution using 30° wide classes.
(ii) Plot the frequency distribution on a suitable type of histogram.
(iii) Is there an overall trend for this data?

7.9 In Section 7.3 it was stated that

$$s^2 = \frac{1}{N}\left(\sum_{i=1}^{N}(w_i - \bar{w})^2\right) \tag{7.4}$$

can be rearranged to give

$$s^2 = \overline{w^2} - (\bar{w})^2 \tag{7.5}$$

Verify this by:
(i) Multiply out $(w_i - \bar{w})^2$ in Eqn. 7.4
(ii) Split the result into three separate summations using the following relationship:

$$\sum (a + b + c + \ldots) = \sum a + \sum b + \sum c + \ldots$$

(iii) Simplify each of the resulting summations using the following result:

$$\sum ka = k \sum a \quad \text{where } k \text{ is a constant.}$$

Table 7.11 The percentage weight of calcium carbonate and organic carbon in the top 100 cm of core RAMA 44PC. Data from Keigwin, L. Jones, G. and Froelich, P. (1992). A 15 000 year palaeoenvironmental record from Meiji Seamount, far north-western Pacific. *EPLS*, **111**, 425–40.

Calcium carbonate (%)	Organic carbon (%)
6.10	0.35
5.30	0.27
5.30	0.28
6.70	0.35
9.00	0.42
7.20	0.43
3.20	0.22
14.30	0.39
13.40	0.48
15.30	0.68
9.00	0.70
3.40	0.87
6.30	0.86
10.20	0.95
10.50	0.95
13.40	1.23
15.10	1.22
5.70	1.25
1.90	1.05
2.00	0.98

(iv) Finally, use the definition of mean and mean square to simplify further and obtain the required result.

7.10 Skewness has been mentioned several times in this chapter and is a measure of the symmetry of a distribution. One of several possible definitions is

$$Skew = (\Sigma \, (w_i - \bar{w})^3)/Ns^3$$

which will equal zero if and only if the distribution is symmetric.
Evaluate this expression for the same 10 pebbles you used in question 7.3.

7.11 Table 7.11 lists the calcium carbonate and organic carbon weight percentages obtained at various points within the top 100 cm of a core from a seamount in the north-western Pacific Ocean.
(i) Calculate a linear regression for this data.
(ii) Plot the resulting best fit straight line together with a scatter plot of the original data (i.e. plotted as individual points not joined together).
(iii) How well do you think a linear regression works in this case?

7.12 The amounts of subsidence at two nearby locations in the south Pyrenean foreland during the mid-Eocene were as follows:

Age (Ma)	Subsidence at Puig d'Olena (m)	Subsidence at Tona (m)
46	12	15
42.5	183	102
40.1	340	381
38.4	491	641
35.6	742	788

Calculate, using linear regression, the best fit gradient and intercept for a plot of subsidence at Tona versus subsidence at Puig d'Olena.

7.13 The SiO_2 contents for samples taken from two adjacent locations in the Yilgarn Craton of Western Australia were:

Mount Monger	Emu
65.08	60.40
67.51	66.50
59.52	70.43
61.60	61.27
63.09	64.07
63.49	66.33
64.31	67.10
62.70	63.88
67.49	66.72
	69.38

Calculate the mean, standard deviation, standard error and 95% confidence limits for the SiO_2 contents at these two locations. Is there a significant difference between the mean SiO_2 content at Mount Monger and that at Emu?

Data from Smithies, R. and Champion, D. (1999). Late Archaean felsic alkaline igneous rocks in the Eastern Goldfields, Yilgarn Craton, Western Australia: a result of lower crustal delamination? *J Geol Soc* **156**, 561–76.

7.14 Use spreadsheet *Bfit.xls* to check your answers to 7.5, 7.11 and 7.12.

7.15 Use spreadsheet *Bfit.xls* to fit the temperature versus depth data given below. Try fitting it with $n = 0, 1, 2, 3$ and 4.

z (km)	T (°C)
100	1150
400	1500
700	1900
2 800	3700
5 100	4300
6 360	4300
7 620	4300
9 920	3700
12 020	1900
12 320	1500
12 620	1150

8 Differential calculus

8.1 Introduction

Calculus, discovered in the seventeenth century by Newton and independently by Leibniz, is where advanced mathematics is usually assumed to begin. Despite this, the mathematical manipulations used in elementary calculus are very simple to apply. The real difficulty with calculus is not in 'How is it done?' but rather in 'How is it used for solving a particular problem?' At first sight a further difficulty is the strange new notation that is required. The mathematical expressions used in calculus are quite different from those you will have encountered previously and this is a significant barrier to understanding. In other words, calculus calculations are easy to do once you understand their strange appearance, but this knowledge is of limited use until you have a good grasp of how to use mathematics more generally for solving real problems. Hopefully, the preceding chapters will have improved your ability to apply simple mathematics. In this chapter, and the next, I shall introduce you to the techniques and notation of calculus and show a few examples of ways to use these techniques. First, however, I should discuss what calculus is for.

8.2 Rates of geological processes

Geology is a process-centred subject. Virtually all geological observations can, and should, be explained by the interaction of various physical, chemical or biological processes. Increasing burial, for example, is a process which has the effect of compressing, heating and chemically altering a sediment. Biological activity such as burrowing is a process which has the effect of mixing the uppermost metre or so of sediment. Convergence of two tectonic plates is a process which can lead to the formation of mountain chains. The point that I wish to make is that all of these processes, and their results, occur at a certain rate. Burial, for example, might occur at a rate of one metre per thousand years or so. Biological mixing could lead to, say, 10% of the upper layer being reworked every year. Plate convergence rates are generally several centimetres per year. These are all examples of rates expressed in terms of the amount of change produced in a certain amount of time, e.g. 1 metre of extra sediment in 1000 years, 10% mixing in 1 year or 3 cm of plate convergence in 1 year.

Compaction of rocks during burial could be expressed in terms of loss of porosity per year (e.g. porosity might reduce by 0.1% per year in a particular case). However, in many cases it is more useful to express compaction in terms of the loss of porosity produced by a given amount of burial. A typical figure might be a 12% loss in porosity produced by 1 km of burial. In this case, the rate of change of porosity is expressed in terms of the amount of change produced in a certain distance.

The key concept here is that we are considering the amount of change in a quantity (e.g. porosity) produced by change in another quantity (e.g. depth of burial).

The difficulty is that, for most processes, this rate of change is not a constant. There might be a loss of 12% porosity as a result of burial by 1 km but further burial down to 2 km might only result in the loss of an additional 5% porosity. In this example, the rate of loss of porosity is faster near the surface than it is at depth. Returning to the near surface mixing caused by biological activity, this is likely to occur at a faster rate in the summer than it does in the winter, so we again have a process whose rate is not a constant.

To avoid this difficulty, rates of change are normally given in terms of their **instantaneous rate**. For example, the rate of loss of porosity at the surface might be 40% per km. This means that the porosity would be reduced by 40% after burial by 1 km if this rate remained constant down to 1 km. In practice, the rate will probably drop off with depth so that the porosity loss after burial by 1 km will be much less than 40%. A more familiar example is the speed of an automobile. If you are travelling at 50 km per hour this means that you would move 50 km in the next hour if you continued at the same rate. It does not mean that you will necessarily move 50 km in the next hour. You might get stopped by heavy traffic after 5 minutes.

Calculus is a set of mathematical tools which allow us to deal with processes whose rates are not constants. Since this is the case for most real processes, calculus is an essential part of all applied mathematics. In particular, **differential calculus** is the tool used to determine the instantaneous rate at which a given process takes place.

8.3 Graphical determination of rates of change

Before we go on to use differential calculus, it is instructive to consider simpler alternatives. This will allow me to introduce some important concepts and is also a relatively simple way to introduce the notation needed for calculus.

In Fig. 8.1 I use the porosity, ϕ, versus depth, z, example once more. In the particular example shown, the rate at which the porosity is falling is very fast at the beginning and becomes less fast at greater depths. This is shown on the graph by the fact that the curve is very steep on the left and becomes much

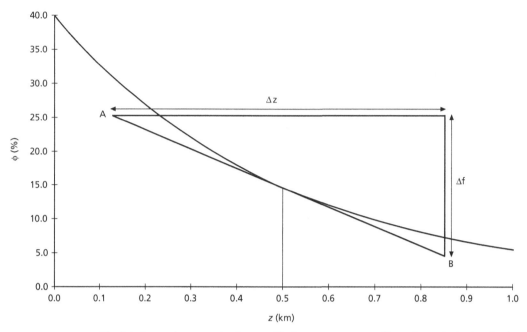

Fig. 8.1 A porosity versus depth curve. The instantaneous rate of loss of porosity is given by the gradient of the tangent.

flatter towards the right. Thus, the porosity drops by about 7% in the first 100 m of burial but drops by about 1% when buried by 100 m starting from a depth of 900 m. Now, what is the instantaneous rate of porosity loss at a depth of 500 m? To determine this, a **tangent** to the curve at $z = 500$ m is drawn as shown. A tangent is a straight line which touches the curve at one, and only one, point. The rate of change that we wish to find is simply the gradient of this tangent.

Question 8.1 The above porosity/depth curve is a smooth graph drawn through the points:

Depth (km)	Porosity (%)
0.0	40.0
0.2	26.8
0.4	18.0
0.6	12.0
0.8	8.1
1.0	5.4

Plot these points on a large sheet of graph paper. Draw a smooth curve through the points and measure the gradient of a tangent to the curve at a depth of 500 m.

Table 8.1 The porosity and gradient data for points taken from Fig. 8.1.	Depth (km)	Porosity (%)	Gradient (%/km)
	0.0	40.0	−80.0
	0.1	32.7	−65.5
	0.2	26.8	−53.6
	0.3	22.0	−43.9
	0.4	18.0	−35.9
	0.5	14.7	−29.4
	0.6	12.0	−24.1
	0.7	9.9	−19.7
	0.8	8.1	−16.2
	0.9	6.6	−13.2
	1.0	5.4	−10.8

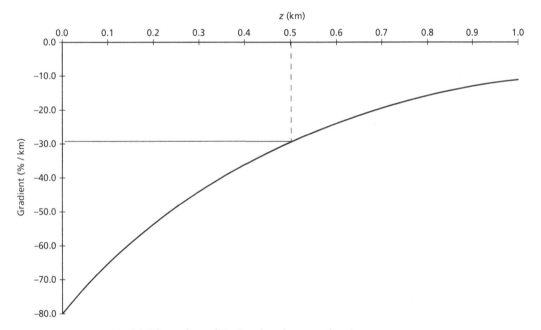

Fig. 8.2 The gradient of Fig. 8.1 plotted as a new function.

Now, having determined the gradient at one particular depth, the gradient at several other depths could also be found. I have done this for the graph shown in Fig. 8.1 and the result is given in Table 8.1. These gradient values can themselves be plotted on a graph as a function of depth. The result is Fig. 8.2. The function in Fig. 8.2 is called the **derivative** of the function in Fig. 8.1. The important point is that the gradient of a curve is another curve rather than a single number.

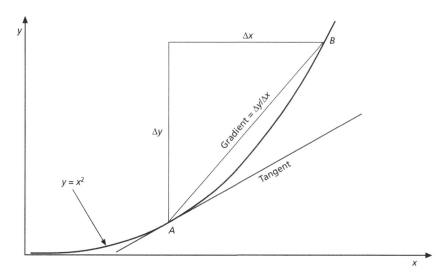

Fig. 8.3 Algebraic determination of the gradient for $y = x^2$. The gradient of line AB would be closer to that of the tangent if B was closer to A.

8.4 Algebraic determination of the derivative of $y = x^2$

The problem with using such graphical approaches is that they are slow and not very accurate. It would be better if there was a more rigorous method for determining the derivative of a given function. One approach would be to use a little algebra to try and calculate the gradient of the tangent. This technique is illustrated in Fig. 8.3 for the case of the function

$$y = x^2 \tag{8.1}$$

Figure 8.3 shows this function and the tangent whose gradient is to be estimated together with the gradient of a line joining two points on the curve. The key points here are that we can attempt to estimate the gradient of the tangent by finding the gradient between two points on the curve, and that the closer the two points are the better the estimate becomes.

Now, since B is considered to be close to A, we can think of its x-location as being a small amount (Δx) greater than the x-location for A. Similarly, the y-location of B is a small amount (Δy) greater than the y-location of A. A slightly different way to state this is that point A is at the location (x, y) whilst point B is at location $(x + \Delta x, y + \Delta y)$. Using this notation, the gradient of the line AB is just

$$\text{Gradient} = \Delta y / \Delta x \tag{8.2}$$

Now, since point B is also on the curve given by Eqn. 8.1 it follows that

$$y + \Delta y = (x + \Delta x)^2 \tag{8.3}$$

All we need to do, to get the gradient of the line AB, is rearrange Eqn. 8.3 to give $\Delta y/\Delta x$. Multiplying out the right-hand side yields

$$y + \Delta y = x^2 + 2x.\Delta x + (\Delta x)^2 \tag{8.4}$$

However, by Eqn. 8.1, we may subtract y from the left-hand side and, simultaneously, remove x^2 from the right-hand side since they are equal. This gives

$$\Delta y = 2x.\Delta x + (\Delta x)^2 \tag{8.5}$$

Finally, dividing Eqn. 8.5 by Δx, leads to

$$\frac{\Delta y}{\Delta x} = 2x + \Delta x \tag{8.6}$$

which is the gradient of line AB.

However, we wish to have the gradient of the tangent. This is obtained by imagining that point B becomes **infinitesimally** close to point A. This means that there is an unimaginably small distance between A and B but they are not quite touching. Under these conditions Δx is said to **tend** towards zero and the gradient of AB tends to the gradient of the tangent at A. As Δx tends towards zero, Eqn. 8.6 becomes

$$\frac{dy}{dx} = 2x \tag{8.7}$$

in which the symbol dy/dx is used to denote the gradient of the tangent and Δx has disappeared from the right-hand side since it has gone to zero.

Now that we have a formula for the gradient we can use it to determine the slope at any point. For example, at $x = 10$

$$\begin{aligned} y &= x^2 \\ &= 10^2 = 100 \end{aligned} \tag{8.8}$$

by Eqn. 8.1 and

$$\begin{aligned} \text{gradient} &= dy/dx \\ &= 2x \\ &= 2 \times 10 = 20 \end{aligned} \tag{8.9}$$

by Eqn. 8.7.

Question 8.2 Using the above equation determine the slope of $y = x^2$ at the points $x = 2$ and $x = 1000$.

Question 8.3 A graph of $y = x^3$ has a gradient given by $dy/dx = 3x^2$. Demonstrate this by following through similar steps to those above.

The main point to understand from the above derivation is that calculus obtains its results by considering what happens in the case of infinitesimally small differences between quantities. This is why it is called **differential calculus** and why the above process is called **differentiation**.

Spreadsheet *Differentiate.xls* shows how the principles of this section work for any function of the form $y = x^n$. The spreadsheet will calculate the gradient of a line joining any two points on such a curve and also gives the gradient of the tangent at the first of the two given points. You should find that, as the distance between the two points becomes smaller, these two gradients become more similar.

8.5 Standard forms

In practice it is never necessary to go through derivations, such as those in Section 8.4, to obtain the derivative of a given function. This is because mathematicians have already done this for the vast majority of cases that geologists will ever use. The results are summarized in tables of derivatives and it is only necessary to go and look these up. The most common examples of these **standard forms** are given in Table 8.2.

Let's go through some of these just to see if they make sense. Firstly we have that if

$$y = x^n \tag{8.10}$$

then

$$\frac{dy}{dx} = nx^{n-1} \tag{8.11}$$

The simplest way to understand this is to use the example followed in Section 8.4. In that section I considered the case of

$$y = x^2 \tag{8.1}$$

This is a particular example of the more general case given by Eqns. 8.10 and 8.11. Comparing Eqn. 8.1 with Eqn. 8.10 you should be able to see that, if $n = 2$, these two are the same. Now, if you replace n by 2 in Eqn. 8.11 you should get

$$\frac{dy}{dx} = 2.x^1$$
$$= 2x \tag{8.12}$$

which is the same as the result given by Eqn. 8.7.

Question 8.4 Using Eqns. 8.10 and 8.11, confirm that you obtained the correct result in question 8.3.

There are two special cases of Eqns. 8.10 and 8.11 that should be considered. Firstly, if $n = 1$, we would have

$$y = x \qquad (8.13)$$

and

$$\begin{aligned} dy/dx &= 1.x^0 \\ &= 1 \end{aligned} \qquad (8.14)$$

since $x^0 = 1$ for any value of x. This is entirely reasonable since, going back to Chapter 2, you should be able to see that Eqn. 8.13 is a straight line through the origin with a gradient of 1.

Secondly, if $n = 0$, we would have

$$\begin{aligned} y &= x^0 \\ &= 1 \end{aligned} \qquad (8.15)$$

and

$$\begin{aligned} dy/dx &= 0.x^{-1} \\ &= 0 \end{aligned} \qquad (8.16)$$

since any number multiplied by zero is also zero. Again this is entirely reasonable since Eqn. 8.15 states that $y = 1$ for all values of x. Thus a graph of y against x would be a horizontal straight line passing through $y = 1$ and this would have a gradient of zero. The important point to remember here is that the derivative of a constant is always zero.

Moving on to other standard forms in Table 8.2, the next result states that

$$\alpha = \sin(\theta) \qquad (8.17)$$

has a gradient given by

$$d\alpha/d\theta = \cos(\theta) \qquad (8.18)$$

Note that, since the variables here are α and θ rather than y and x, we must differentiate α with respect to θ to give a gradient in the form $d\alpha/d\theta$. The reasonableness of the result given by Eqns. 8.17 and 8.18 should become clear if you look at Fig. 8.4 in which the sine function and its derivative are shown as a function of angle in radians. Note that the sine function begins with a large positive gradient, becomes flat at $\pi/2$ radians, has a large negative slope at π radians, becomes flat again at $3\pi/2$ radians and finally becomes steeper and positive again as a complete rotation is performed at 2π radians.

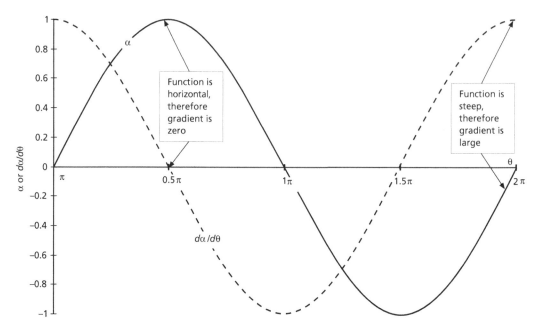

Fig. 8.4 The sine function and its gradient. Note that, at 0 radians, the function is steep and so the gradient is high, at $\pi/2$ radians, the function is horizontal and so the gradient is zero etc.

y	dy/dx
x^n	nx^{n-1}
$\sin(x)$	$\cos(x)$
$\cos(x)$	$-\sin(x)$
$\tan(x)$	$1/\cos^2(x)$
e^x	e^x
$\ln(x)$	$1/x$

Table 8.2 The most common standard forms for differentiation.

This function certainly looks very like the cosine function, as predicted by Eqn. 8.18, but is it really exactly the same? The answer is yes, provided the sine function is given as a function of the angle in radians. Thus, whenever trigonometric ratios are used in calculus, it should always be assumed that angles are measured in radians rather than degrees.

Skipping over the other trigonometric cases in Table 8.2, the exponential case is particularly interesting. Figure 8.5 should help to remind you what functions of the type

$$y = a^x \tag{8.19}$$

look like if a is a positive constant. The point is that these functions are always positive and increase consistently as you move to increasing values

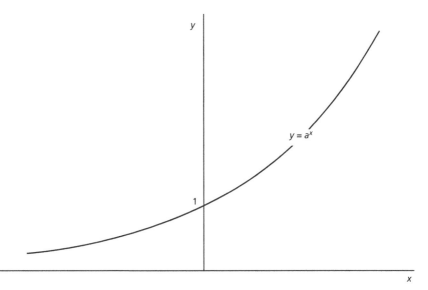

$y = a^x$

Fig. 8.5 The function $y = a^x$. Note that both the function and its slope increase continuously from left to right.

of x. They also always have a positive gradient which becomes increasingly steep with increasing x. It follows from this that the derivative of a function like Eqn. 8.19 has a very similar appearance to the original function. In fact the number e (= 2.71 . . .), and only this number, has the very special property that, if

$$y = e^x \qquad (8.20)$$

then

$$dy/dx = e^x \qquad (8.21)$$

Question 8.5 Find the derivatives of the following functions. Be careful to get the notation correct.
(i) $y = x^{20}$ Differentiate y with respect to x.
(ii) $w = e^z$ Differentiate w with respect to z.
(iii) $w = x^z$ Differentiate w with respect to x, assuming z is constant.

8.6 More complicated expressions

The trouble with standard forms is that real problems are rarely that simple. In practice, the types of expression that need to be differentiated do not appear in Table 8.2. For example, if we wished to differentiate

$$y = e^x + x^{20} - \sin(x) \qquad (8.22)$$

how would we do it? The problem is that, whilst we now know how to find the derivative of each of the individual terms, how do we differentiate the whole expression? The answer is very simple indeed. All we do is differentiate the terms individually and add or subtract the results as appropriate. Thus, Eqn. 8.22 gives

$$dy/dx = e^x + 20x^{19} - \cos(x) \tag{8.23}$$

since the derivative of e^x is e^x, the derivative of x^{20} is $20x^{19}$ and the derivative of $\sin(x)$ is $\cos(x)$. The next problem is that an expression might be multiplied by a constant. How would you differentiate

$$\alpha = 32 \cos(\theta)? \tag{8.24}$$

Now, the cosine function is in Table 8.2 so we can find its derivative $(= -\sin(\theta))$. The function $32 \cos(\theta)$ is simply 32 times steeper than the function $\cos(\theta)$ and therefore

$$d\alpha/d\theta = -32 \sin(\theta) \tag{8.25}$$

In other words, if a function is multiplied by a constant, simply multiply the derivative by the same constant.

Question 8.6 Differentiate the following:
(i) $y = x^{10} + \tan(x)$, with respect to x;
(ii) $y = x^2 + x^3 + x^4$, with respect to x;
(iii) $w = 75e^z$, with respect to z;
(iv) $\alpha = 10 \sin(\theta) + 13 \cos(\theta)$, with respect to θ.

The rules for addition and for multiplication by a constant enable us to differentiate something useful at last. What is the rate of increase in temperature with depth at a depth of 1000 km? We can start with one of the temperature equations introduced in Chapter 2. The temperature, T, in degrees Celsius at a depth, z, in km is given approximately by

$$T = (-8.255 \times 10^{-5})z^2 + 1.05z + 1110 \tag{2.6}$$

Now, if we want to know the gradient at a particular depth, we must first obtain an expression for gradient as a function of depth. This is simply done by differentiating Eqn. 2.6 with respect to z. This yields

$$\begin{aligned} dT/dz &= 2(-8.255 \times 10^{-5})z + 1.05 \\ &= (-1.651 \times 10^{-4})z + 1.05 \end{aligned} \tag{8.26}$$

To obtain the gradient at $z = 1000$ km we now substitute this depth into Eqn. 8.26 which gives

$dT/dz = (-1.651 \times 10^{-4} \times 10^3) + 1.05$
$\qquad = -0.1651 + 1.05$
$\qquad = 0.88°C \ km^{-1}$ (8.27)

Thus, at a depth of 1000 km, the Earth's temperature increases by 0.88°C for each additional km of depth. Note that, to solve problems of this type, you must differentiate first and then substitute the depth. You cannot solve this problem by first substituting the depth of 1000 km into Eqn. 2.6 and then differentiating afterwards.

Question 8.7 Equation 2.8,

$T = az^4 + bz^3 + cz^2 + dz + e$ (2.8)

with values for the constants of

$a = -1.12 \times 10^{-12}$
$b = 2.85 \times 10^{-8}$
$c = -0.000 \ 310$
$d = 1.64$
$e = 930$

gave a slightly different temperature versus depth function. Use this equation to evaluate the gradient at a depth of 1000 km. Compare to the answer which was obtained above using Eqn. 2.6.

8.7 The product rule and the quotient rule

The next complication that must be dealt with is the case of a function which is the product of two simpler functions. For example

$y = x^2.\sin(x)$ (8.28)

which is the product of the function x^2 with the function $\sin(x)$. We need a general rule for dealing with these cases so, instead of the specific problem of Eqn. 8.28 I will use the more general expression

$y = u.v$ (8.29)

where u and v are any two functions. For example, in Eqn. 8.28

$u = x^2$ (8.30)

and

$v = \sin(x)$ (8.31)

The product rule states that

$$dy/dx = u.(dv/dx) + v.(du/dx) \qquad (8.32)$$

Thus, there are two terms to the answer. The first term consists of the first function multiplied by the derivative of the second whilst the second term consists of the second function times the derivative of the first.

More formally, from Eqns. 8.30 and 8.31 it follows that the derivative of the first function is

$$du/dx = 2x \qquad (8.33)$$

and the derivative of the second function is

$$dv/dx = \cos(x) \qquad (8.34)$$

Substituting Eqns. 8.30, 8.31, 8.33 and 8.34 into Eqn. 8.32 yields

$$dy/dx = x^2.\cos(x) + 2x.\sin(x) \qquad (8.35)$$

Question 8.8 Find the derivatives of
(i) $\alpha = x^2.e^x$
(ii) $y = 3w^2.\sin(w)$
(iii) $z = x.\cos(x) + x^3.\tan(x)$
(iv) $B = 3\sigma^4.\ln(\sigma) + 17\sigma^2$

Question 8.9 The **quotient rule** states that if

$$y = u/v$$

where u and v are functions of x, the derivative of y with respect to x is given by

$$\frac{dy}{dx} = \frac{v(du/dx) - u(dv/dx)}{v^2}$$

Use this rule to find the derivative of $y = x^4/\sin(x)$
(Hint: Let $u = x^4$ and $v = \sin(x)$)

8.8 The chain rule

Sometimes it is necessary to make a **substitution** to find the derivative of an expression. If this is done then the **chain rule** should be used to obtain the required solution. Take the case of

$$y = \sin(x^2) \qquad (8.36)$$

To differentiate this, make the substitution

$$z = x^2 \qquad (8.37)$$

which gives

$$y = \sin(z) \tag{8.38}$$

Now, it is a simple matter to differentiate Eqn. 8.38 using the table of standard forms. The result is

$$dy/dz = \cos(z) \tag{8.39}$$

However, note that this is the derivative of y with respect to z. If we could differentiate Eqn. 8.36 the result would be the derivative of y with respect to x. So, how do we change from one to the other? The chain rule must be used. This states that

$$dy/dx = (dy/dz).(dz/dx) \tag{8.40}$$

for the case of the symbols we have used in this example. The left-hand side of this is the derivative we require. The right-hand side has two parts, the first of which is the derivative we have found (i.e. Eqn. 8.39). To complete the calculation, therefore, we need to know dz/dx. This can easily be determined by differentiating Eqn. 8.37. Thus,

$$dz/dx = 2x \tag{8.41}$$

which, after substituting into Eqn. 8.40, gives

$$\begin{aligned} dy/dx &= \cos(z).2x \\ &= \cos(x^2).2x \\ &= 2x.\cos(x^2) \end{aligned} \tag{8.42}$$

With the chain rule available to us, we can find the gradient of the porosity/depth function introduced in Chapter 2, i.e.

$$\phi = \phi_0\, e^{-z/\lambda} \tag{2.17}$$

where ϕ is porosity, z is depth and ϕ_0 and λ are constants. To find the gradient of porosity with depth, first make the substitution

$$x = -z/\lambda \tag{8.43}$$

This leads to

$$\phi = \phi_0 e^x \tag{8.44}$$

giving

$$d\phi/dx = \phi_0 e^x \tag{8.45}$$

In addition, Eqn. 8.43 gives

$$dx/dz = -1/\lambda \tag{8.46}$$

Finally, the chain rule for this problem is

$$d\phi/dz = (d\phi/dx).(dx/dz) \tag{8.47}$$

This is constructed simply by writing down, on the left-hand side, the derivative we would like to know and then writing on the right-hand side the two derivatives we have found (i.e. Eqns. 8.45 and 8.46). The result can be checked simply by imagining that dx in the two parts of the right-hand side is 'cancelled'. The result is $d\phi/dz$ which is the same as the left-hand side. If this doesn't work then the chain rule has been incorrectly written down for your problem. (N.B. Although this is a good way to check the correctness of the chain rule, it must be emphasised that this 'cancellation' is a convenient fiction. Functions such as $d\phi/dx$ do not denote $d\phi$ divided by dx, they denote the gradient of ϕ as a function of x. In other words, $d\phi/dx$ is a single function, it is not one function divided by another.)

Thus, we can now substitute Eqns. 8.45 and 8.46 into the chain rule (Eqn. 8.47) to give

$$d\phi/dz = (\phi_0 \, e^x).(-1/\lambda)$$
$$= (-\phi_0/\lambda).e^{-z/\lambda} \tag{8.48}$$

A simpler form for this expression arises by noting that Eqn. 8.48 is very similar to Eqn. 2.17 leading to

$$d\phi/dz = -\phi/\lambda \tag{8.49}$$

For example, if the surface porosity of a shale is 60% and the decay distance λ is 500 m, then the porosity and porosity gradient at a depth of 1000 m are

$$\phi = 60.\exp(-1000/500)$$
$$= 8.12\% \tag{8.50}$$

from Eqn. 2.17, and

$$d\phi/dz = -8.12/500$$
$$= -0.0162\% \, m^{-1} \tag{8.51}$$

from Eqn. 8.49.

Question 8.10 Differentiate $x = \ln(y^2)$ with respect to y.

8.9 So, what's it for?

Calculus is probably one of the most important topics in the whole of applied maths. Some of the most useful applications for calculus need **differential equations** (beyond the scope of this book) or **integral calculus** (the subject of the next chapter). Nevertheless, there are useful things that can be done with simple differential calculus. One of these follows from the fact that the derivative of a function gives the rate of change of the function. Thus, the amount by which the function changes over a given distance can easily be found. An example used in geophysics is essential in **gravity surveying**.

Gravity surveying techniques are based upon the fact that the strength of gravity at the Earth's surface is increased by dense material in the subsurface and reduced by less dense materials. Hence, in principle, measurements of the strength of gravity can indicate the subsurface distribution of rocks of differing densities. However, altitude also affects the strength of gravity. Gravity is reduced as distance from the centre of the Earth increases. Now, if a gravity survey is made over a mountain range, each measurement would be taken at a different altitude. Any difference in the gravity would then be due to the combination of changes in the subsurface density distribution and changes in altitude. However, the geophysical survey is intended to reveal only the subsurface density variation since there are easier ways of determining altitude! Thus, before interpreting the gravity survey, it is normal to remove the altitude effect. This is called the **free-air correction**. Note that there are several more corrections to gravity measurements which should also be made but the details of these are beyond the scope of this book. However, the free-air correction is probably the most important correction which must be made to gravity data before it is interpreted.

So, what is this correction? We start with the formula for the acceleration due to gravity introduced in Chapter 3.

$$g = GM/r^2$$
$$= GMr^{-2} \tag{3.6}$$

The next step is to differentiate Eqn. 3.6 with respect to distance, r, from the Earth's centre. This gives

$$dg/dr = -2GMr^{-3} \tag{8.52}$$

Substituting

r = 6370 km = 6.37×10^6 m
G = 6.672×10^{-11} m³ kg⁻¹ s⁻²
M = 5.95×10^{24} kg

then gives

$$dg/dr = -3.072 \times 10^{-6} \text{ s}^{-2} \tag{8.53}$$

In other words, the strength of gravity reduces by 3.072×10^{-6} m s⁻² for every metre of increased height. For example, if two measurements are separated by an altitude of 350 m, there will be a difference of 350 × 3.072×10^{-6} = 0.00 108 m s⁻² entirely caused by the altitude difference. This correction should be added to the measurement with the higher altitude before a comparison is made.

Question 8.11 A gravity survey gave the following results:

Measurement	Gravity (m s^{-2})	Height above datum (m)
1	9.78 350	0
2	9.78 012	1100
3	9.78 196	500

Using the altitude correction given in Eqn. 8.53 determine whether or not the differences in these measurements are significant assuming that the measurements are accurate to within 0.00 002 m s^{-2}.

8.10 Higher derivatives

The whole point about differentiation is that it produces a new function which is the gradient of the initial function. Obviously, it is quite possible to differentiate the result again to produce the gradient of the gradient function (see Fig. 8.6). Figure 8.6 shows a simple function, y, which is differentiated to give the gradient function also shown in Fig. 8.6. Finally, the result of differentiating the gradient function itself is shown as the gradient of the gradient. Following line A through the three functions in the figure should make this clear. Line A shows the values of the three functions at around $x = -4.5$. The function y has a positive gradient here since it slopes up towards the right and so the gradient function is also positive. However, the gradient function slopes downwards towards the right and therefore the gradient of the gradient is negative.

The only thing that is remotely difficult about forming these **higher derivatives** is the notation. This is best explained using a simple example. Take the case of

$$\chi = \sin(\psi) \tag{8.54}$$

Differentiation produces

$$d\chi/d\psi = \cos(\psi) \tag{8.55}$$

The derivative of this new function is then found by differentiating again. The derivative of cosine is −sine and therefore further differentiation of this result gives

$$d^2\chi/d\psi^2 = -\sin(\psi) \tag{8.56}$$

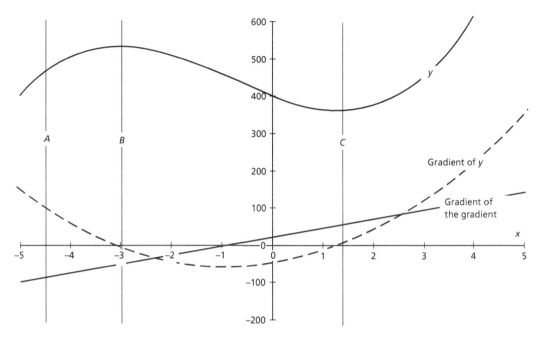

Fig. 8.6 A simple function together with its gradient (first derivative) and the gradient of its gradient (second derivative).

which is called the second derivative of Eqn. 8.54. Note that the fact that this result has been obtained by differentiating the function twice is indicated simply by placing superscripted 2s as indicated on the left-hand side. This process can be carried on indefinitely. For example, differentiation of −sine yields −cosine and therefore Eqn. 8.56, after further differentiation, gives the third derivative

$$d^3\chi/d\psi^3 = -\cos(\psi) \tag{8.57}$$

where the superscripted 3s are introduced since the original function has now been differentiated three times.

Question 8.12 If $y = 3x^2 + 2x - 1$, obtain dy/dx and d^2y/dx^2. Also write down the third and fourth derivatives making sure to get the notation correct.

8.11 Maxima and minima

The higher derivatives can be used to find the **maxima** and **minima** of functions. Two such points are shown on Fig. 8.6. Lines B and C pass through points on y which are a maximum (all adjacent points are lower) and a minimum (all adjacent points are higher), respectively. Now, at these points the gradient is zero. Thus, lines B and C indicate points where the derivative is zero. Note also that, to the left of the maximum, the gradient is positive and to the right it is negative. Thus, the gradient at this point is decreasing and the gradient of the gradient is therefore negative in this area. In summary, the second derivative must be negative at points corresponding to maxima. Conversely, at the point corresponding to the minimum in y, the second derivative must be positive.

To summarize:

1 At maxima and minima the first derivative is zero;
2 At maxima the second derivative is negative;
3 At minima the second derivative is positive.

This ability to determine the maxima and minima of a function allows a large class of new problems to be solved using differential calculus. These are called **optimization** problems and apply whenever we wish to determine the optimum value of some parameter. An example follows that might appear a little complicated. However, each of the separate steps has been covered above and you should be able to follow these steps with occasional glances at the preceding sections.

Large faults in the subsurface typically occur at an angle of around 60° to the horizontal. One theory about why this might be so is because this angle minimizes the amount of work that has to be done on the crust in order to achieve a given amount of extension. To test this theory we need to find an expression for the amount of work done as a function of fault dip. We then find the angle for which this function is a minimum. In other words, we find the optimum angle from the point of view of minimizing work done. Figure 8.7 shows the essential elements of this problem.

The amount of work necessary to achieve this extension will obviously increase if the fault length increases since there is then a larger surface for friction to act on. In addition, the amount of work done will also increase as the amount of movement on the fault increases. Therefore, inspection of Fig. 8.7 shows that, if the fault is shallow, a lot of work must be done because the fault is long. On the other hand, Fig. 8.7 also shows that, if the fault is steep, a lot of work must be done because a lot of movement along the fault is needed. We need to find an intermediate angle where these two effects are both relatively small.

The simplest assumptions that we can make about this problem are that the work expended, w, is proportional to the length of the fault, L, and also proportional to the amount of fault movement, m. Thus,

$$w = \alpha L m \qquad (8.58)$$

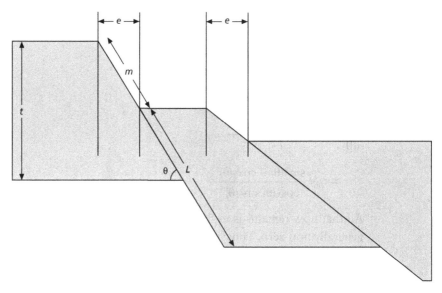

Fig. 8.7 The amount of fault movement, m, and the fault length, L, for a given thickness, t, and extension, e. Both m and L depend upon the fault dip.

where α is a positive constant whose precise value is unimportant in what follows. Expressions for L and m in terms of fault dip are now needed. These follow from simple trigonometry which gives

$$L = t/\sin(\theta) \tag{8.59}$$

and

$$m = e/\cos(\theta) \tag{8.60}$$

where t is the thickness of the brittle crust and e is the amount of extension. Thus, combining Eqns. 8.58 to 8.60 gives

$$w = \alpha et/[\cos(\theta)\sin(\theta)] \tag{8.61}$$

This is our expression for work done as a function of fault dip. All that remains is to find the angle θ for which w is minimized. From the earlier discussion, this occurs where the derivative of Eqn. 8.61 is zero. The derivative can be found using the quotient rule introduced in Section 8.7. For this example, use the substitutions

$$u = \alpha et \tag{8.62}$$

and

$$v = \cos(\theta)\sin(\theta) \tag{8.63}$$

from which it follows that

$$du/d\theta = 0 \qquad\qquad (8.64)$$

and

$$dv/d\theta = \cos^2(\theta) - \sin^2(\theta) \quad \text{(from product rule)} \qquad (8.65)$$

The quotient rule then gives

$$\frac{dw}{d\theta} = \frac{v.du/d\theta - u.dv/d\theta}{v^2}$$

$$= \alpha et \frac{\sin^2(\theta) - \cos^2(\theta)}{\cos^2(\theta)\,\sin^2(\theta)} \qquad\qquad (8.66)$$

All that now remains is to find the angles for which this is zero. Now αet is generally non-zero. Thus, if

$$dw/d\theta = 0.0 \qquad\qquad (8.67)$$

then

$$[\sin^2(\theta) - \cos^2(\theta)] = 0.0$$

i.e.

$$\sin^2(\theta) = \cos^2(\theta) \qquad\qquad (8.68)$$

One solution to Eqn. 8.68 is

$$\theta = 45° \qquad\qquad (8.69)$$

We must now determine whether this is a maximum or a minimum. Remember that both maxima and minima have a derivative of zero. However, we specifically need a minimum since we are looking for a fault dip which minimizes the work needed for extension. To test whether a fault dip of 45° produces a minimum we must obtain the second derivative of w. This is done by differentiating Eqn. 8.66 which produces

$$\frac{d^2w}{d\theta^2} = \frac{2\alpha et(\sin^4\theta + \cos^4\theta)}{\cos^3\theta\sin^3\theta} \qquad\qquad (8.70)$$

Now, if the fault dip is 45°, this gives a value of

$$\frac{d^2w}{d\theta^2} = 8\alpha et \qquad\qquad (8.71)$$

which is positive since α, e and t are all positive (proof of Eqns. 8.70 and 8.71 is given as an exercise later). A fault dip of 45° therefore corresponds to a minimum in w. Thus, the theory predicts that faults should occur at 45° which contradicts the observation that faults are generally considerably steeper than this.

Thus, we have a simple quantitative test of a theory for fault formation. Since observations and theory do not agree here, we are forced to conclude that something is missing in our theory. The important point is that the preceding mathematics has forced us to conclude that the simple fault model is inadequate. It would be very difficult, in the absence of such an analysis, to decide whether or not the simple fault dip theory was correct. The next step would be to consider whether the model can be modified to make it more accurate or whether it should be completely abandoned. However, more complete theories of fault formation are beyond the scope of this book.

8.12 Further questions

8.13 The thickness, S, of post-rift sediment in a simple extensional basin with a good sediment supply is well approximated by

$$S = S_{max}(1 - e^{-t/\tau})$$

where S_{max} and τ are constants and t is time since rifting ceased.
(i) Give an expression for the sediment thickness immediately after rifting (i.e. at $t = 0$).
(ii) Give an expression for the sediment thickness when t is very large.
(iii) Give an expression for the rate of sedimentation immediately after rifting.
(iv) What is the rate of sedimentation when t is very large?
(v) If $S_{max} = 3$ km and $\tau = 50$ My, what is the numerical value of the sedimentation rate immediately after rifting?

8.14 Figure 8.8 illustrates the barred basin model of evaporite formation. Sea water of normal salinity floods through the inlet of the basin and evaporation within the basin increases this salinity to the point where halite and

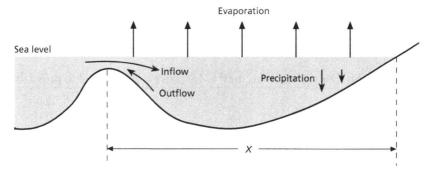

Fig. 8.8 Barred basin model of evaporite formation. Sea water enters through the inlet and evaporation results in increasing salinity shoreward of the inlet. Eventually the water is saline enough for precipitation of halite and other evaporites to occur.

other evaporites precipitate from the water and are deposited on the basin floor.

A simple mathematical theory for this process predicts that the salinity, s, within the basin increases with distance, x, from the inlet according to

$$s = s_0 \alpha X / (\alpha X - x)$$

where s_0 is the salinity of the sea water at the inlet, α is a constant and X is the basin width.

(i) If water at the shoreline (i.e. at $x = X$) has three times the salinity of the sea water at the inlet, what is the value of α?

(ii) With this value for α and a sea water salinity of 30 ppm, determine the salinity at a point half way across the basin (i.e. at $x = X/2$).

(iii) Prove that the rate of increase in salinity with distance is given by

$$ds/dx = s/(\alpha X - x)$$

(iv) Evaluate this salinity gradient at a point half way across the basin if its width is 10 000 m. Hence, estimate the salinity at points 1 km on either side of the basin centre.

8.15 In Section 8.11, it was stated that if

$$\frac{dw}{d\theta} = \alpha et \, \frac{\sin^2\theta - \cos^2\theta}{\cos^2\theta \sin^2\theta} \tag{8.66}$$

then

$$\frac{d^2w}{d\theta^2} = \frac{2\alpha et(\sin^4\theta + \cos^4\theta)}{\cos^3\theta \sin^3\theta} \tag{8.70}$$

Prove this result using the following steps:

(i) Use the chain rule to differentiate $u = 1/\sin^2\theta$ and $v = 1/\cos^2\theta$.

(ii) Use these results to differentiate Eqn. 8.66 rewritten in the form

$$dw/d\theta = \alpha et[(1/\cos^2\theta) - (1/\sin^2\theta)]$$

Simplify the result.

(iii) Use the fact that, for $\theta = 45°$, $\cos\theta = \sin\theta = 1/\sqrt{2}$ to prove that

$$\frac{d^2w}{d\theta^2} = 8\alpha et$$

at this angle.

8.16 The thickness of a bottomset bed at the foot of a delta can often be well approximated by the expression

$$t = t_o \exp(-x/X)$$

where t is thickness, x is distance from the bottomset bed start and t_o and X are constants.

(i) Determine an expression for the rate of change of bed thickness.

(ii) Evaluate this expression at a point 3 km from the bottomset bed start if $X = 5$ km and $t_o = 10$ m.

8.17 Equation 1.2 gives the relationship between burial depth, z, and sediment age, a, in a lake bed as

$$a = kz \tag{1.2}$$

where k is a constant.

(i) Differentiate this with respect to z to give the rate of change of age with depth (i.e. the change in age produced by change in depth by 1 m).

(ii) Hence, if the depth is in error by Δz metres, write down an expression for the error, Δa, in the age.

(iii) For $k = 3000$ y m^{-1} and a depth error of 0.1 m, what is the age error?

8.18 The height, h, of a cliff can be determined from the angle, α, to the cliff top measured from a point at a distance, x, from the cliff base. From elementary trigonometry (i.e. Chapter 5) it follows that the cliff height is given by

$$h = x \tan(\alpha)$$

How high is the cliff if a measurement taken at a distance of 100 m gave an angle of 20°? Using the same approach as that for question 8.17, determine the error in cliff height if this angle was in error by 2°.

9 Integral calculus

9.1 Introduction

Integration is a mathematical procedure for finding the sum of an infinite number of infinitely small quantities. Figure 9.1 shows a simple example where the problem is to calculate the area of the grey shaded region. Figure 9.1 also shows a way in which an approximate answer might be found. The area under the curve is slightly less than the total area of the three rectangles shown. A more accurate answer could obviously be obtained by using a larger number of thinner rectangles and the correct answer should result if an infinite number of infinitesimally thin rectangles were used (i.e. integration). This chapter is concerned with how such integration is done and with its applications.

Integration may seem a little abstract and not particularly useful. Nothing could be further from the truth. As I will show later, integration allows such things as the variation in density with depth in the Earth to be estimated, the heat generated in the crust to be determined and the approximate volume of Mount Fuji to be calculated. However, I'll start by solving the problem posed

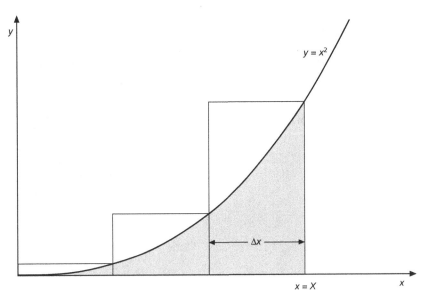

Fig. 9.1 Estimating the area under the curve $y = x^2$. The area of the three rectangles is easily found but this is slightly more than the grey area. A better approximation could be obtained by using more, thinner rectangles.

by Fig. 9.1 and this will lead to the very important result that integration is simply the inverse procedure to differentiation.

9.2 The area under $y = x^2$

Returning to Fig. 9.1, the problem is to find the area under the curve $y = x^2$ between $x = 0$ and $x = X$. The area of one of the rectangles is $y.\Delta x$ where y is its height and Δx its width. Hence, the area of n such rectangles is

$$A = \sum_{i=1}^{n} y_i \Delta x$$

$$= \sum_{i=1}^{n} x_i^2 \Delta x \tag{9.1}$$

where y_i is the height of the ith rectangle and x_i is the x-location at which the rectangle touches the curve. If the rectangles are constructed as shown in Fig. 9.1 then $x_i = i\,\Delta x$. Hence, the area can be rewritten as

$$A = \sum_{i=1}^{n} i^2 \Delta x^3$$

$$= \Delta x^3 \sum_{i=1}^{n} i^2 \tag{9.2}$$

The summation in Eqn. 9.2 has the solution (not proved here)

$$\sum_{i=1}^{n} i^2 = \frac{n(n+1)(2n+1)}{6} \tag{9.3}$$

For example, if $n = 3$, the sum is $1^2 + 2^2 + 3^2 = 14$ and the right-hand side of Eqn. 9.3 becomes $(3 \times 4 \times 7)/6 = 14$. However, if n is very large, Eqn. 9.3 simplifies since $n + 1$ is very close to n under these circumstances and, similarly, $2n + 1 \approx 2n$. This gives

$$\sum_{i=1}^{n} i^2 \approx \frac{n.n.2n}{6} = \frac{n^3}{3} \text{ for } n \text{ large.} \tag{9.4}$$

Substituting Eqn. 9.4 into 9.2 and noting that $X = n.\Delta x$, then leads to

$$A = \Delta x^3 \sum_{i=1}^{n} i^2$$

$$= \Delta x^3 \frac{n^3}{3}$$

$$= \frac{X^3}{3} \tag{9.5}$$

which is the answer we require. Don't worry too much if you find the details of the above derivation a little confusing. The important point is that you understand the principle of taking the area of an infinite number of infinitely small rectangles. Using the result (i.e. Eqn. 9.5) is very straightforward now. For example, the area under the curve $y = x^2$ for $X = 2$ simply $A = 2^3/3 = 8/3 = 2.667$.

Now, in a similar manner to that discussed for differentiation, when the limiting case of an infinite number of infinitely small rectangles is taken, the notation has to change slightly to indicate that this has been done. Equation 9.1 changes as follows:

$$A = \sum_{i=1}^{n} x_i^2 \Delta x \tag{9.1}$$

becomes

$$A = \int_0^X x^2 \, dx \tag{9.6}$$

where Δx has become dx (cf. differential calculus) and the finite sum symbol Σ has been replaced by the $\int$ symbol (called an integration sign) indicating an infinite sum. Finally, the 0 below the integration sign and the X above it indicate that the area has been found between $x = 0$ and $x = X$. The 0 and the X are called the **limits of integration**. Incidentally, the integration sign itself is derived from the old-fashioned 'long s' (e.g. the King James Bible of 1611 was printed with sentences such as, 'For this cauſe left I thee in Crete, that thou ſhouldeſt ſet in order the things that are wanting,' Titus 1, v. 5). and is used because integration is a special form of summation.

Finally, compare the result from this section that

$$\int_0^X x^2 \, dx = \frac{X^3}{3} \tag{9.7}$$

to the fact that, from Chapter 8,

$$\text{if } y = \frac{x^3}{3} \text{ then } \frac{dy}{dx} = x^2 \tag{9.8}$$

Close inspection of Eqns. 9.7 and 9.8 should show a relationship between these two results. This link will be developed in the following sections and will, eventually, allow us to perform integrations in a much simpler way than that given above.

9.3 Indefinite integration

Before a simpler integration method can be presented, we'll have to digress a little to discuss a related, but slightly different, procedure called **indefinite integration**. Indefinite integration is the reverse process to differentiation. For example, if

$$y = \sin(\theta) \tag{9.9}$$

then

$$dy/d\theta = \cos(\theta) \tag{9.10}$$

from which it follows that

$$\int \cos(\theta)\, d\theta = \sin(\theta) \tag{9.11}$$

or, in words, the indefinite integral of $\cos(\theta)$ is $\sin(\theta)$ because differentiating $\sin(\theta)$ produces $\cos(\theta)$. However, the derivative of $\sin(\theta) + 3$ is also $\cos(\theta)$ and, in general, the derivative of $\sin(\theta) + k$ equals $\cos(\theta)$ for any constant k. Thus, the general solution to the integral of $\cos(\theta)$, is $\sin(\theta) + k$ where k is an unknown constant. Equation 9.11 is therefore not quite correct and should be rewritten

$$\int \cos(\theta)\, d\theta = \sin(\theta) + k \tag{9.12}$$

Note that, in this indefinite integration, integration limits are not given.

Similarly, one of the problems in Chapter 8 involved differentiating

$$y = e^x + x^{20} - \sin(x) \tag{8.22}$$

to give

$$dy/dx = e^x + 20x^{19} - \cos(x) \tag{8.23}$$

From this it follows that

$$\int (e^x + 20x^{19} - \cos(x))\, dx = e^x + x^{20} - \sin(x) + k \tag{9.13}$$

Equation 9.13 is entirely equivalent to Eqns. 8.22 and 8.23. This is just a different way of writing almost the same thing. Note that the integration is performed with respect to θ in Eqn. 9.12 and with respect to x in Eqn. 9.13. To indicate this, Eqn. 9.12 contains a $d\theta$ term whilst Eqn. 9.13 contains a dx term.

y	$\int y \, dy$
nx^{n-1}	$x^n + k$
$\cos(x)$	$\sin(x) + k$
$-\sin(x)$	$\cos(x) + k$
$1/\cos^2(x)$	$\tan(x) + k$
e^x	$e^x + k$
$1/x$	$\ln(x) + k$

Table 9.1 A table of standard integrals formed by swapping around the columns of Table 8.2.

> **Question 9.1** From the last chapter it should be clear that the derivative of $5e^{2\alpha}$ with respect to α is $10e^{2\alpha}$. Using this fact, write an integral equation similar to Eqns. 9.12 or 9.13. Don't forget the constant and remember that you are now integrating with respect to α and thus the function to be integrated is enclosed between $\int$ and $d\alpha$.

From the preceding discussion it should be clear that a table of standard integrals can be produced from a table of standard derivatives simply by swapping around the columns. Thus, reproducing Table 8.2 and rearranging the columns gives a table of integrals (Table 9.1) in which k is an integration constant. Note that the first and third examples in Table 9.1 are not particularly simple as they stand, and therefore of limited use as **standard integrals**. The first expression in Table 9.1 can be improved upon by starting with the fact that if

$$y = \frac{x^{n+1}}{n + 1} \tag{9.14}$$

then

$$\frac{dy}{dx} = x^n \tag{9.15}$$

from which it follows that

$$\int x^n \, dx = \frac{x^{n+1}}{n + 1} + k \tag{9.16}$$

However, if $n = -1$ this expression does not give a sensible answer (try it!). In this particular case use the result from Table 9.1 that $\int x^{-1} dx = \ln(x) + k$.

Similarly, it is easy to show that

$$\int \sin(x) \, dx = -\cos(x) + k \tag{9.17}$$

Table 9.2 A table of standard integrals formed by simplifying some of the results in Table 9.1.

y	$\int y \, dy$
x^n	$\dfrac{x^{n+1}}{n+1} + k$
$\cos(x)$	$\sin(x) + k$
$\sin(x)$	$-\cos(x) + k$
$1/\cos^2(x)$	$\tan(x) + k$
e^x	$e^x + k$
$1/x$	$\ln(x) + k$

Using these two results in place of the first and third in Table 9.1 gives a table of standard integrals (Table 9.2). This list is a very short one. There are large books containing standard integrals for many other functions.

Question 9.2 Using Table 9.2 evaluate:

(i) $\int \cos(x) \, dx$

(ii) $\int \xi^{10} \, d\xi$ (Hint: This is solved using the first result in Table 9.2.)

9.4 Definite integration

We are now in a position to show the link between indefinite integration discussed in Section 9.3 and the 'infinite sum of infinitely small quantities' type of integration discussed in Section 9.2. The method discussed in Section 9.3 was called indefinite integration because the unknown constant results in an answer which is not completely determined. In contrast, the integration discussed in Section 9.2 produces an unambiguous answer and is therefore called **definite integration**.

However, indefinite and definite integration are very closely linked. The simplest way to show this is with an example such as

$$y = \int_1^2 x \, dx \tag{9.18}$$

The first step is to ignore the integration limits and proceed as for indefinite integration to give

$$y = \left[\frac{x^2}{2} + k \right]_1^2 \tag{9.19}$$

where the limits have been written outside the square brackets just to remind us that they are still there. So what do we do with them? The answer is to substitute each in turn in place of x and to subtract the results. Thus

$$y = \left(\frac{2^2}{2} + k\right) - \left(\frac{1^2}{2} - k\right)$$
$$= 2 - 0.5 + k - k$$
$$= 1.5 \qquad (9.20)$$

which is the final answer. Notice that the constant, k, cancels out and therefore, in definite integration only, it can be ignored. Also note that the lower limit result is subtracted from the upper limit result.

Question 9.3 Use the technique described above to confirm the solution, from Section 9.2, that

$$\int_0^X x^2 \, dx = \frac{X^3}{3}$$

Question 9.4 Derive a general expression for the area under the curve $y = x^n$ between $x = 0$ and $x = X$. Check your results for specific values of n and X by using spreadsheet *Integ.xls*.

A trigonometric example is also worth going through:

$$\int_{\pi/2}^{3\pi/2} \cos(a) \, da = [\sin(a)]_{\pi/2}^{3\pi/2}$$
$$= \sin(3\pi/2) - \sin(\pi/2)$$
$$= -1.0 - 1.0$$
$$= -2.0 \qquad (9.21)$$

Note that this answer is negative. Negative areas are those which lie below the horizontal axis. In fact, both cos and sin have an area of zero when integrated from 0 to 2π radians (i.e. 0° to 360°) since these functions lie half above the axis (positive area) and half below (negative area). Look back to Fig. 5.9 to make this clear.

9.5 Integration of more complex expressions

As with differentiation, the problem with standard forms is that real problems are rarely that simple. We need a few rules for dealing with more complex cases. Some of these are similar to those for differentiation. For example, if a function is multiplied by a constant, the integral of that function is multiplied by the same constant, i.e.

$$\int a.f(x)\, dx = a \int f(x)\, dx \qquad (9.22)$$

where a is a constant and $f(x)$ is any function of x. For instance

$$\int 3\sin(x)\, dx = 3 \int \sin(x)\, dx$$
$$= 3(-\cos(x) + k)$$
$$= -3\cos(x) + c \qquad (9.23)$$

where c is an integration constant (which happens to equal $3k$ but this is unimportant since k itself is not known).

Question 9.5 Evaluate the following definite integral:

$$y = \int_0^1 4e^x\, dx$$

Another rule is that the integral of the sum of several functions is simply given by the sum of the integrals, i.e.

$$\int [f(x) + g(x) - h(x)]\, dx = \int f(x)\, dx + \int g(x)\, dx - \int h(x)\, dx \qquad (9.24)$$

where $f(x)$, $g(x)$ and $h(x)$ are any functions of x. Note that, although three functions are shown in Eqn. 9.24, this principle applies to any number of functions. An example of this is

$$\int [\cos(x) - x^2]\, dx = \int \cos(x)\, dx - \int x^2\, dx$$
$$= \sin(x) - (x^3/3) + k \qquad (9.25)$$

Equations 9.22 and 9.24 can be combined to solve problems such as

$$\int [e^\theta + 5\theta^{19}]\, d\theta = \int e^\theta\, d\theta + 5 \int \theta^{19}\, d\theta$$
$$= e^\theta + 5(\theta^{20}/20) + k$$
$$= e^\theta + (\theta^{20}/4) + k \qquad (9.26)$$

> **Question 9.6** Using the above rules and Table 9.2, integrate $(10/x) + \cos(x)$ with respect to x.

These are all the integration rules that we need in this book, although you will find textbooks full of many more rules and tips. Integration is, in practice, frequently very difficult or even impossible. However, with the additional rules you can find in other texts and the published lists of standard forms, many problems can be solved.

9.6 Simple applications of integration

Integration can be used to solve many problems. For example, how would you go about measuring the volume of Mount Fuji? One way to do this is to split the mountain into a series of thin slices (Fig. 9.2), each of which has a volume which can be calculated. Now, the volume of a slice is simply its area times its thickness, Δz. I have used Mount Fuji for simplicity since, being a very symmetric volcanic structure, the slices are in fact well approximated by small circular discs. Thus, since the area of the disc in Fig. 9.2 is πr^2 (where r is the disc radius), the disc volume, ΔV, is given by

$$\Delta V = \pi r^2.\Delta z \tag{9.27}$$

The volume of the entire mountain is then approximately equal to the sum of the volumes of all these slices, i.e.

$$V = \sum_{i=1}^{N} \Delta V_i \tag{9.28}$$

$$= \sum_{i=1}^{N} \pi r_i^2.\Delta z \tag{9.29}$$

where ΔV_i and r_i are the volume and radius, respectively, of the ith disc and there are N discs from the mountain base to its peak.

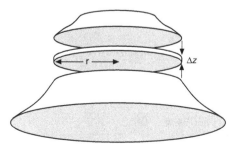

Fig. 9.2 Estimating the volume of Mount Fuji by summing the volume of lots of thin slices.

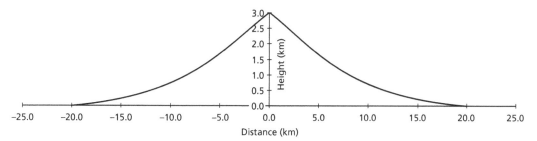

Fig. 9.3 The shape of Mount Fuji modelled using Eqn. 9.31.

So, how many discs do we need? A small number of thick discs would produce a mountain whose sides were made of large steps. As the discs become thinner, and their number greater, the sides would become gradually smoother. Thus, as the discs become thinner, more discs are needed but the result becomes more accurate. To get a good result we should use a very large number of very thin discs. To get the best possible result, we should use an infinite number of infinitely thin discs. If this is done then Δz is replaced by an infinitely thin dz and, to indicate that N has become infinite, the summation sign is replaced by an integral sign (cf. the transformation of Eqn. 9.1 to Eqn. 9.6 in Section 9.2). Equation 9.29 then becomes

$$V = \int_{Z_{min}}^{Z_{max}} \pi r^2 .dz \tag{9.30}$$

The Z_{min} and Z_{max} indicate that the first disc is at the mountain base and the last is at the mountain top. Equation 9.30 is an example definite integration interpreted as finding the sum of an infinite number of infinitely small quantities as discussed in Section 9.2.

To solve Eqn. 9.30 we require a definite form for how the radius varies with height up the mountain. From a map of Mount Fuji it can be shown that, to a good approximation,

$$r^2 = \frac{400z}{3} - \frac{800\sqrt{z}}{\sqrt{3}} + 400 \text{ km}^2 \tag{9.31}$$

where z goes from 0 at the base to 3 km at the top. Thus, at the mountain base, $z = 0$ and Eqn. 9.31 gives a radius of 20 km. At the mountain top, $z = 3$ and Eqn. 9.31 gives a radius of zero. Figure 9.3 shows the form of this expression for predicting the mountain shape. Substituting Eqn. 9.31 into Eqn. 9.30 then gives

$$V = \int_0^3 \pi \left[\frac{400z}{3} - \frac{800\sqrt{z}}{\sqrt{3}} + 400 \right] dz$$

$$= \pi \int_0^3 \frac{400z}{3} dz - \pi \int_0^3 \frac{800\sqrt{z}}{\sqrt{3}} dz + \pi \int_0^3 400 \, dz$$

$$= \pi \left[\frac{400z^2}{6} - \frac{800z^{1.5}}{1.5\sqrt{3}} + 400z \right]_0^3$$

$$= \pi(600 - 1600 + 1200)$$
$$= 200\pi = 628 \text{ km}^3 \tag{9.32}$$

which is the approximate volume of Mount Fuji. From this, we could estimate the age of Mount Fuji if we knew the rate of production of volcanic material. Alternatively, the average rate of production could be estimated if some other measure of volcano age could be found.

Question 9.7 The thickness of a sedimentary body varies as shown below

Distance (m)	Thickness (m)
0.0	0.0
100.0	1.0
200.0	2.0
400.0	3.0
1000.0	2.5
1500.0	2.0
1800.0	1.0
2000.0	0.0

Use spreadsheet *Bfit.xls* to find a best fit polynomial through this data. Hence, find the cross-sectional area.

One final example is worth going through to show the breadth of topics integration is applicable to. In structural geology there are several different ways of measuring strain (i.e. the amount of deformation). One such quantity is called the strain, s, and is given by

$$s = \Delta L/L_i \tag{9.33}$$

where L_i is the initial length of a strained material and ΔL is the amount by which the deformation increases the length (see Fig. 9.4).

An alternative is to consider the final strain as resulting from a large number of small deformation episodes. In the limit, the deformation can be

Fig. 9.4 Stretching material from an initial length of L_i to a final length of L_f.

Question 9.8 Heat is generated by radioactivity in the crust at a rate measured in μW m^{-3} or, occasionally, kW km^{-3} (Note that these two units are actually the same, i.e. 1 μW m^{-3} = 1 kW km^{-3}). For example, each cubic kilometre of continental upper crust generates about 1 kW of heat (which is about the same as a typical electric room heater). This heat generation rate actually decreases with depth until, below the crust, hardly any heat is generated in this way.

If, in a particular piece of crust 30 km thick, the heat generation rate, Q, can be approximated by

$$Q = y/20 \text{ kW km}^{-3} \qquad (9.34)$$

where y is the distance from the base of the crust:
(i) Determine the heat generation rate at 0, 10, 20 and 30 km from the crust base. Sketch a graph of the result.
(ii) Write down an expression for the heat generated in a small box-shaped volume Δz km thick whose other dimensions are 1 km by 1 km. (Hint: What is the volume of such a box?)
(iii) Give an expression for the approximate heat generated by a column of such boxes by writing an equation similar to Eqn. 9.29.
(iv) Write down the expression that results from allowing the thickness of these crust volumes to tend to zero.
(v) Evaluate the resulting equation which gives the rate of flow of heat from each square kilometre at the surface.
(vi) What area would be needed to generate as much heat as a small-sized power station (say 100 MW)?

---

thought of as resulting from an infinite number of infinitely small extensions. In each episode the length is increased by an amount dL and there are an infinite number of these such that the length increases from an initial length L_i to a final length L_f. Thus, the resulting strain measurement, called the total natural strain, ε, is given by

$$\varepsilon = \int_{L_i}^{L_f} \frac{dL}{L}$$

$$= [\ln(L)] \frac{L_f}{L_i}$$

$$= \ln(L_f/L_i) \tag{9.35}$$

This can be compared to a third method of quantifying strain called the stretch, S, which is simply the final length divided by the initial length. From this it follows that

$$\varepsilon = \ln(S) \tag{9.36}$$

i.e. the total natural strain is simply the natural logarithm of the stretch.

9.7 Integrating discontinuous functions

The difficulty with real problems is that they are often not well approximated by the simple functions considered up to this point. In particular, especially in geological problems, there are often sharp discontinuities where the properties of interest change dramatically. A good example is the density profile of the Earth (Fig. 9.5) which has sharp changes at various depths such as the mantle/core boundary.

This profile has been determined using a large number of different geophysical measurements. In particular, one important constraint on possible density profiles is that they must be consistent with the known mass of the

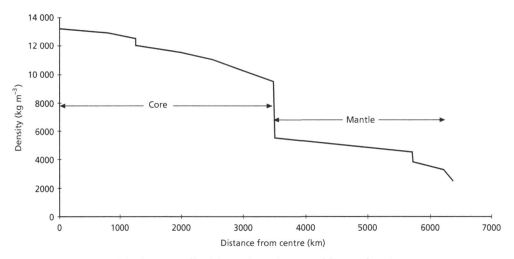

Fig. 9.5 The density profile of the Earth as a function of distance from its centre.

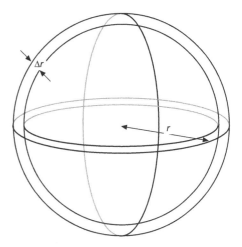

Fig. 9.6 A thin spherical shell of radius r and thickness Δr.

Earth (see Chapter 3 on how this is known). If the average density of the proposed profile is too high then the mass will also be too high and vice versa. The total mass is found by first considering the mass of a thin spherical shell of thickness Δr and radius r (Fig. 9.6). The volume of a thin shell is approximately its surface area times its thickness. Thus, since the surface area of a sphere is $4\pi r^2$, the volume is $4\pi r^2 \Delta r$ and the mass is $4\pi r^2 \rho \Delta r$ where ρ is the shell density. This leads to a total mass estimate of

$$M = \sum_{i=1}^{N} 4\pi r_i^2 \rho_i \Delta r \tag{9.37}$$

where there are N shells in total and r_i is the radius and ρ_i is the density of the ith shell. The accuracy of this expression increases as the number of shells increases and the shell thickness decreases until, in the limit of infinitely many infinitely thin shells, it becomes the integral

$$M = \int_{0}^{R} 4\pi r^2 \rho \, dr \tag{9.38}$$

where r is the radius of the Earth.

All we need now is an expression for how density varies with radius. For simplicity we can approximate the density by a single constant value for the core and a single constant value for the mantle. Let's assume densities of 11 000 kg m^{-3} and 4500 kg m^{-3}, respectively. How do we handle the change-over at the core/mantle boundary radius of 3480 km? Simple, we integrate using a density of 11 000 kg m^{-3} from the centre to the boundary to give the mass of the core. A separate calculation using a density of 4500 kg m^{-3} can then be used to integrate from the boundary to the surface to give the mantle mass. The total mass is then just the sum of the two. Thus

$$M = \int_0^{3480} 4\pi r^2 . 11\,000\ dr + \int_{3480}^{6371} 4\pi r^2 . 4500\ dr$$

$$= 44\,000\pi[r^3/3]_0^{3480} + 18\,000\pi r^3/3]_{3480}^{6371}$$

$$= (46\,077 \times (3.48 \times 10^6)^3) + 18\,850((6.371 \times 10^6)^3 - (3.48 \times 10^6)^3)$$

$$= (1.94 \times 10^{24}) + (4.08 \times 10^{24})$$

$$= 6.02 \times 10^{24}\ \text{kg} \tag{9.39}$$

which compares quite well to the mass calculated in Chapter 3 of 5.95×10^{24} kg. The discrepancy would disappear if a density function closer to that shown in Fig. 9.5 were used instead of the simple two density model used above.

9.8 Further questions

9.9 The Earth has the shape of a slightly flattened sphere, i.e. the polar radius is slightly less than the equatorial radius. Figure 9.7 is an exaggeration of this to illustrate the point. The precise shape is one in which sections cut parallel to the equator are circular whilst sections through the poles are elliptical. Such a shape is called an ellipsoid.

In an ellipsoid, the radius, r, of a circle of constant latitude at a distance z from the equator is given by:

$$r^2 = r_e^2(1 - (z^2/r_p^2)) \tag{9.40}$$

(i) What is the volume of the disc illustrated in the figure in terms of its radius, r, and thickness, Δz?
(ii) Write an expression for the approximate volume of the Earth by assuming it is filled with a stack of N such discs.
(iii) Rewrite the expression for the case of an infinite number of infinitely thin discs. Evaluate the resulting integral. (Hint: You will need to integrate

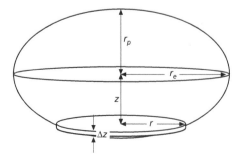

Fig. 9.7 An exaggerated depiction of the Earth as an ellipsoid. The polar radius, r_p, is less than the equatorial radius, r_e. A thin disc parallel to the equator is also shown which is a distance z from the equator and which has a radius r.

from $z = -r_p$ to $z = r_p$. The result will be an algebraic expression rather than a simple number.)

(iv) If the equatorial radius of the Earth is 6378 km and the polar radius is 6357 km, what is the volume?

(v) How does this compare to the volume for a sphere of radius equal to the average of the polar and equatorial values?

9.10 The thickness of a bottomset bed at the foot of a delta can often be well approximated by the expression

$$t = t_0 \exp(-x/X)$$

where t is thickness, x is distance from the bottomset bed start and t_0 and X are constants.

(i) Imagine approximating this sedimentary bed in cross section by a series of rectangles of height t_i and width Δx. What is the area of each rectangle?

(ii) Now write down an approximate expression for the cross-sectional area of the entire bottomset bed.

(iii) By considering the limiting case of an infinite number of infinitesimally thick rectangles, write down and evaluate an integral equation giving the total cross-sectional area.

(iv) If the present-day rate of sediment supply is 10 m^2 yr^{-1}, $X = 5$ km and $t_0 = 1$ m, estimate the time taken to form the bed assuming the sediment supply rate has not altered through time.

Appendix A: useful equations

1 $\log_y(y^x) = x$

2 If $y = ax^2 + bx + c$, then $y = 0$ when $x = \dfrac{-b \pm \sqrt{b^2 - 4ac}}{2a}$

3 For the following triangle (in which upper case letters are angles and lower case letters are lengths):

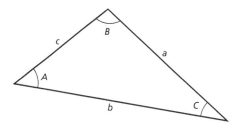

180° rule: $A + B + C = 180°$

Sine rule: $\dfrac{\sin(A)}{a} = \dfrac{\sin(B)}{b} = \dfrac{\sin(C)}{c}$

$$a^2 = b^2 + c^2 - 2bc \, \cos(A)$$

Cosine rule: $b^2 = a^2 + c^2 - 2ac \, \cos(B)$

$$c^2 = a^2 + b^2 - 2ab \, \cos(C)$$

4 For N measurements, w_i, $i = 1, 2, 3, \ldots, N$

average or mean, $\bar{w} = \dfrac{1}{N} \sum_{i=1}^{N} w_i$

sample variance, $as^2 = \dfrac{1}{N} \left(\sum_{i=1}^{N} (w_i - \bar{w})^2 \right)$

$$= \overline{w^2} - \bar{w}^2$$

unbiased estimate of population variance, σ^2 is given by, $\hat{s}^2 = \dfrac{N}{N-1} s^2$

5 The normal, or Gaussian, distribution is:

$$P(x) = \frac{\exp\left[-(x - \bar{x})^2/2\sigma^2\right]}{\sigma\sqrt{2\pi}}$$

6 For N measurement pairs x_i, y_i, $i = 1, 2, 3, \ldots, N$, the best fit straight line (linear regression), of form $y = mx + c$, is given by:

$$m = \frac{n \sum xy - \sum x \sum y}{n \sum x^2 - (\sum x)^2}$$

and $c = \bar{y} - m\bar{x}$

in which all summations are performed over the N measurement pairs, i.e.

$\sum xy$ is shorthand for $\displaystyle\sum_{i=1}^{N} x_i y_i$ etc.

7 Standard derivatives:

y	dy/dx
x^n	nx^{n-1}
$\sin(x)$	$\cos(x)$
$\cos(x)$	$-\sin(x)$
$\tan(x)$	$1/\cos^2(x)$
e^x	e^x
$\ln(x)$	$1/x$

8 Product rule. If $y = u.v$, where u and v are functions of x, then

$$\frac{dy}{dx} = u.\frac{dy}{dx} + v.\frac{du}{dx}$$

9 Quotient rule. If $y = u/v$, where u and v are functions of x, then

$$\frac{dy}{dx} = \frac{v.(du/dx) - u.(dy/dx)}{v^2}$$

10 Chain rule. If, in an expression $y = y(x)$, the substitution $u = u(x)$ is made, then

$$\frac{dy}{dx} = \frac{dy}{du}.\frac{du}{dx}$$

11 Standard integrals:

y	$\int y\, dx$
x^n	$x^{n+1}/(n+1) + k$
$\sin(x)$	$-\cos(x) + k$
$\cos(x)$	$\sin(x) + k$
$1/\cos^2(x)$	$\tan(x) + k$
e^x	$e^x + k$
$1/x$	$\ln(x) + k$

Appendix B: answers to problems

Chapter 1

1.1 $Age = 1500 \times Depth$; $Depth = 1$ m, $Age = 1500$ years
$\qquad\qquad\qquad\qquad Depth = 2$ m, $Age = 3000$ years
$\qquad\qquad\qquad\qquad Depth = 5.3$ m, $Age = 7950$ years
$\qquad Age = 3000 \times Depth$; $Depth = 1$ m, $Age = 3000$ years
$\qquad\qquad\qquad\qquad Depth = 2$ m, $Age = 6000$ years
$\qquad\qquad\qquad\qquad Depth = 5.3$ m, $Age = 15\,900$ years

1.2 (i) 5^{104}; (ii) 5^{400}; (iii) x^5; (iv) $Depth^5$; (v) $T_0^{12} = 10^{12}$
1.3 (i) 1×10^3; (ii) 2×10^3; (iii) 2.5×10^3; (iv) 2.523×10^3; (v) 2.3×10^7;
(vi) 7×10^9
1.4 (i) 1×10^{-3}; (ii) 2×10^{-3}; (iii) 2.5×10^{-3}; (iv) 2.523×10^{-3};
(v) 2.3×10^{-6}; (vi) 7×10^{-9}
1.5 1000 years $= 3.16 \times 10^{10}$ s
1.6 $0.01\% = 0.01$ parts per hundred $= 0.1$ parts per thousand $= 100$ ppm
1.7 (i) 4×10^{109}; (ii) 2.65×10^{109}; (iii) 1×10^{211}; (iv) 2.35×10^{211}
1.8 (i) 6×10^{300}; (ii) 1.6×10^{221}; (iii) 2; (iv) 2×10^{50}
1.9 5509 kg m^{-3}
1.10 300 years
1.11 8.71×10^{-7}
1.12 Total mass gained $= \Delta M \times A_e = 6 \times 10^5 \times 365.24 \times 4.5 \times 10^9$
$\qquad\qquad\qquad\qquad\qquad = 9.86 \times 10^{17}$ kg
$\qquad$ Fractional gain $= 9.86 \times 10^{17}/5.95 \times 10^{24} = 1.66 \times 10^{-7} = 166$ ppb
1.13 1.08×10^{21} m^3
1.14 (i) $A = w/v_s$; (ii) 1.25×10^8 years $= 125$ million years
1.15 20 km

Chapter 2

2.1 1.025 My
2.2 Gradient $= (1\,050\,000 - 1\,010\,000)/(100 - 20) = 40\,000/80$
$\qquad\qquad\quad = 500$ y m^{-1}
2.3 Rate $= 3800$ y m^{-1}, lake dried out $545\,000$ years ago. N.B. Your gradient should lie within about 500 y m^{-1} of mine and your intercept within about $10\,000$ years of mine

Fig. b1

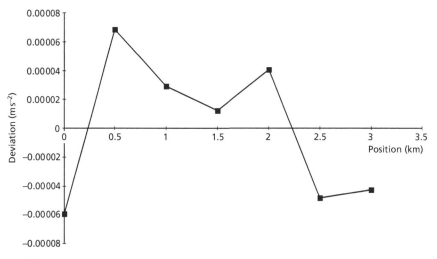

Fig. b2

2.4 $\tau = mP + \tau_0$. The graph should be a straight line of gradient m and intercept τ_0

2.5 $a = 2$, $b = -10$ and $c = 6$

2.6 Equation 2.6 gives 3403°C; Equation 2.8 gives 3648°C; Equation 2.8 is much closer to the true value of 3700°C

2.7 Your answers should be close to 0.45

2.8 $0.6 \times 2^{-2} = 0.6/2^2 = 0.6/4 = 0.15$

2.9 $\phi = 0.7\exp(-1/2) = 0.7 \times 0.6065 = 0.425$

2.10 25

2.11 See Fig. b1: (i) 23 y cm^{-1}; (ii) 1.8 y cm^{-1} (to within about 0.4 cm); (iii) 1030 years

2.12 $C = 200 \times 0.5^{5.5} = 4.42$ ppm

2.13 Graph should be straight line of gradient -10^{-7} and intercept 6.91. Age is 23 million years

2.14 See Fig. b2. Ore body lies between 0.5 km and 2 km

2.15 (i)

Z	p	Z	p
0	3	12	1.646 435
2	2.714 512	14	1.489 756
4	2.456 192	16	1.347 987
6	2.222 455	18	1.219 709
8	2.01 096	20	1.103 638
10	1.819 592		

(ii) p_0 is the accumulation rate in zero water depth. Z is the depth at which accumulation decreases to $p_0/e \sim 1$ m ky^{-1}

Chapter 3

3.1 $k = $ Age / Depth; hence, $k = 3000/3 = 1000$ y m^{-1}

3.2 Age of top = Age $- k$. Depth $= 60\,000 - 5000 \times 10 = 10\,000$ years

3.3 $w/x = [3y/(4z)]/[2y/(4z)] = 12yz/8yz = 12/8 = 1.5$

3.4 (i) $5(x + 2y) = 5x + 10y$; (ii) $5(x + 2.2y) = 5x + 11y$; (iii) $5.5(x + 2y) = 5.5x + 11y$; (iv) $5a(x + 2y) = 5ax + 10ay$; (v) $(x - 2y)(x + 2y) = x^2 - 4y^2$; (vi) $(x + 2y)^2 = x^2 + 4y^2 + 4xy$

3.5 Depth $= (1/k)(\text{Age} - \text{Age of top}) = [(1/k)\ \text{Age}] - [(1/k)\ \text{Age of top}] = (\text{Age}/k) - (\text{Age of top})/k$

3.6 $6ax + 3ay = 3a(2x + y)$

3.7 See Fig. b3

3.8 $x = 1$ for both roots

3.9 2170 km and 10 550 km

3.10 $\rho = \rho_g[1 - (V_p/V)] = M/V$, hence $M = V\rho_g[1 - (V_p/V)]$
$$= \rho_g(V - V_p)$$

and so $\rho_g = M/(V - V_p)$

If $M = 205$ kg, $V = 0.11$ m^3 then average density $= M/V = 205/0.11 = 1864$ kg m^{-3}

For a porosity of 0.32, $V_p = 0.32 \times 0.11 = 0.0352$ m^3

From above $\rho_g = M/(V - V_p) = 205/(0.11 - 0.0352) = 2741$ kg m^{-3}

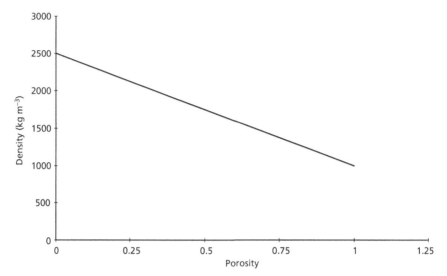

Fig. b3

3.11 $v_1 = \dfrac{2(\rho_p - \rho_f)g r_1^2}{9\eta}$ and $v_2 = \dfrac{2(\rho_p - \rho_f)g r_2^2}{9\eta}$, hence $\dfrac{v_1}{v_2} = \dfrac{r_1^2}{r_1^2} = \left(\dfrac{r_1}{r_2}\right)^2$

If $r_1 = 0.1$ mm and $r_2 = 1$ mm then $v_1 = 0.01v_2$. Hence, smaller particle settles at one-hundredth of the speed of the larger particle. Larger particle therefore takes 0.1 days = 2.4 hours to settle

3.12 (i) $b = -ax - (c/x)$; (ii) $b^2 = a^2x^2 + (c^2/x^2) + 2ac$, therefore $b^2 - 4ac = a^2x^2 + (c^2/x^2) - 2ac$;

(iii) Easiest way is $[ax - (c/x)]^2 = [ax - (c/x)][ax - (c/x)]$
$$= (a^2x^2 - ac) - (ac - c^2/x^2)$$
$$= a^2x^2 + (c^2/x^2) - 2ac;$$

(iv) $b^2 - 4ac = [ax - (c/x)]^2 = (b + 2ax)^2$ (from (i) above)

hence, $b + 2ax = \sqrt{b^2 - 4ac}$ giving $x = \dfrac{-b \pm \sqrt{b^2 - 4ac}}{2a}$ as required

Chapter 4

4.1 15 million years

4.2 11 ppb

4.3 $a = 1.1; b = 2; c = 3$

4.4 Approximation gives $3000 = (-1 \times 10^{-4})z^2 + z + 1000$

Further manipulation leads to $z = \dfrac{-1 \pm \sqrt{0.2}}{-2 \times 10^{-4}}$ approximating further that

$\sqrt{0.2} = 0.5$ then gives $z = 2500$ km or $z = 7500$ km

4.5 $G = \dfrac{r^2 g}{M}$, therefore units$(G) = \dfrac{\text{m}^2 \ \text{m s}^{-2}}{\text{kg}} = \text{m}^3 \ \text{kg}^{-1} \ \text{s}^{-2}$

4.6 $T = 0a + T_0 = T_0$ as expected

4.7 (i) $\ln(t_0) = (x/X) + \ln(t)$

(ii) and (iii) $\ln(t_0) = (1/X) + \ln(5)$ and $\ln(t_0) = (4/X) + \ln(0.1)$
solving simultaneously gives $X = 0.767$ km and $t_0 = 18.4$ m

4.8 and 4.9 $a = -6.93 \times 10^{-13}°\text{C km}^{-4}$, $b = -5.55 \times 10^{-5}°\text{C km}^{-2}$;
$c = 4390°\text{C}$

Predicted central temperature = 4390°C

Predicted surface temperature = 1010°C

4.10 (i) wrong; (ii) wrong; (iii) wrong; (iv) wrong

Chapter 5

5.1 (i) $B = 60°$; $b = 10.1$ cm; $c = 11.5$ cm

(ii) $A = 25°$; $B = 135°$; $c = 2.4$ cm

5.2 Church–transmitter distance = 3.2 km

Church–exposure distance = 4.5 km

Transmitter–exposure distance = 3.5 km

5.3 (i) $180° = \pi$ radians; (ii) $90° = \pi/2$ radians; (iii) $270° = 1.5 \ \pi$ radians;
(iv) $100° = 100\pi/180 = 1.75$ radians

5.4 See Fig. b4. From this $\sin(\theta) = \sqrt{3}/2$; $\cos(\theta) = 1/2$; $\tan(\theta) = \sqrt{3}$ For the
other angle sine and cosine are swapped around and tangent is $1/\sqrt{3}$

5.5 142 m

5.6 22°

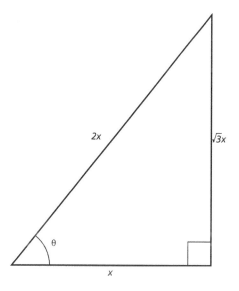

Fig. b4

5.7 (i) $C = 40°$; (ii) $B = 74.6°$; (iii) $a = 2.28$ km

5.8 $a^2 = b^2 + c^2 - 2bc \cos(90) = b^2 + c^2 - 0 = b^2 + c^2$

5.9 22.3° opposite side of length 200 m

108.2° opposite side of length 500 m

49.5° opposite side of length 400 m

5.10 (i) $B = 119.7°$, $C = 20.3°$, $a = 3.70$ km; (ii) $B = 74.6°$, $C = 65.4°$, $c = 2.82$ km; (iii) $C = 80°, b = 4.04$ km, $c = 4.60$ km

5.11 $\tan(\theta_C) = -4/-4 = 1$

$\tan(\theta_D) = -8/8 = -1$

5.12 48.2°

5.13 (i) Scalar; (ii) Vector; (iii) Scalar; (iv) Vector

5.14 See Fig. b5. N.B. Vector $\underline{e}$ has length of zero and is not shown

Fig. b5

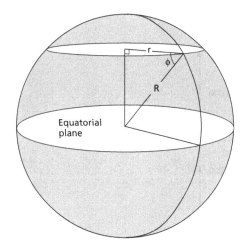

Fig. b6

5.15 $\cos(\theta) = x/a$, therefore $x = a\cos(\theta)$. $\sin(\theta) = y/a$, therefore $y = a\sin(\theta)$

5.16

Vector 1: x-component $= 10\cos(60) = 5.0$ m, z-component $= 10\sin(60)$
$= 8.66$ m

Vector 2: x-component $= 5\cos(65) = 2.11$ m, z-component $= 5\sin(65)$
$= 4.53$ m

Vector 3: x-component $= 12\cos(45) = 8.49$ m, z-component $= 12\sin(45)$
$= 8.49$ m

Vector sum: x-component $= 15.60$ m, z-component $= 21.68$ m

Magnitude $= 26.71$ m. Dip $= 54.3°$

5.17 (i) $\cos(15°) = 0.966$; (ii) $\sin(1.2$ radians$) = 0.932$; (iii) $\tan^{-1}(0.5) =$
$26.6°$; (iv) $\cos^2(27°) = 0.794$; (v) $(\tan(0.5°))^{-1} = 115$

5.18 See Fig. b6. From this $\cos(\phi) = r/R$ giving $r = R\cos(\phi)$

5.19 436 m

5.20 From section, $\tan(\text{apparent dip}) = 150/1000$; from map, true direction
of maximum dip $= 162°$; hence angle between section and dip section
is $32°$; equation 5.32 then gives true dip $= 10.0°$

5.21 352° E of N

5.22 11.3°

5.23 0.277 m s^{-1} at 57.2° E of N

Chapter 6

6.1 (i) See Fig. b7; (ii) See Fig. b8; (iii) See Fig. b9; gradient $= -0.9$,
therefore $b = 0.9$

6.2 Mass $= 55\,000$ kg; foot area $= 2$ m^2

Fig. b7

Fig. b8

Fig. b9

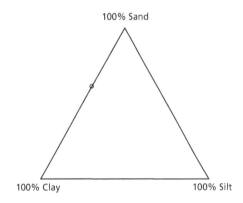

Fig. b10

6.3 See Fig. b10

6.4 See Fig. b11

6.5 10 mm in both cases

6.6 10°–20° = 12.50 mm; 70°–80° = 7.80 mm

6.7 10°–20° = 8.81 mm; 70°–80° = 11.01 mm

6.8 A triangular diagram

6.9 A simple x–y plot

6.10 A polar graph

6.11 (i) A great circle; (ii) Bed dip and pole dip differ by 90°. The dip directions differ by 180°; (iii) See table below and Fig. b12

Fig. b11

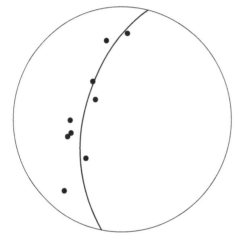

Fig. b12

Pole dip	Pole azimuth
20	362
25	350
65	311
50	250
54	270
20	215
50	220
53	253
53	326

(iv) Fig. b12 also shows a great circle which lies reasonably close to the data which is therefore consistent with a simple fold

6.12 (i) A polar plot of some kind, for example a stereonet; (ii) A triangular diagram; (iii) A log–normal graph with %TOC on the logarithmic axis

Chapter 7

7.1 TTTT, HTTT, THTT, TTHT, TTTH, HHTT, HTHT, HTTH, THHT, THTH, TTHH, THHH, HTHH, HHTH, HHHT, HHHH
Hence,

0 Heads	1 Head	2 Heads	3 Heads	4 Heads
1	4	6	4	1

Therefore 2 heads is the most likely result from 4 coin tosses

7.2 A sample

7.3 (i) mean = 337.6 g; (ii) median = 349 g; (iii) variance = 2838.24 g^2; (iv) population variance = 3153.6 g^2; (v) standard deviation = 56.16 g

7.4 $P_{2.1} = 0.964$; $P_{1.1} = 0.729$; $P_{1.0} = 0.683$; hence, $P_{1.06} = 0.711$. Thus, probability required = $(0.964 - 0.711)/2 = 0.127$ *Gauss.xls* gives 0.125

7.5 Gradient = 1974 y m^{-1}, intercept = 61.8 y

7.6 No, these ratios are indistinguishable within error

7.7 19 degrees of freedom gives t(95) = 2.1 as for the strike data. Hence, Dip(A) = 22.8 ± 1.2° and Dip(B) = 20.3 ± 1.8°

7.8 (i)

Range (°)	Frequency	Probability
0–30	9	0.45
30–60	0	0
60–90	2	0.1
90–120	2	0.1
120–150	0	0
150–180	0	0
180–210	3	0.15
210–240	0	0
240–270	1	0.05
270–300	2	0.1
300–330	0	0
330–360	1	0.05

(ii) See Fig. b13

(iii) Trend is towards NNE

7.9 (i) $s^2 = \dfrac{1}{N} \sum\limits_{i=1}^{N} (w_i^2 + \bar{w}^2 - 2w_i\bar{w})$;

(ii) $s^2 = \dfrac{1}{N} \left(\sum\limits_{i=1}^{N} w_i^2 + \sum\limits_{i=1}^{N} \bar{w}^2 - \sum\limits_{i=1}^{N} 2w_i\bar{w} \right)$;

Fig. b13

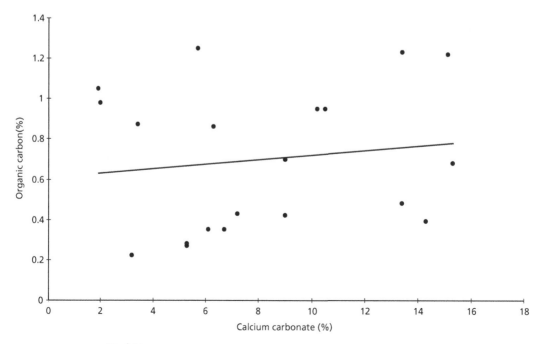

Fig. b14

(iii) $s^2 = \left(\dfrac{1}{N} \sum\limits_{i=1}^{N} w_i{}^2 \right) + \left(\dfrac{1}{N} \bar{w}^2 \sum\limits_{i=1}^{N} 1 \right) - \left(2\bar{w} \dfrac{1}{N} \sum\limits_{i=1}^{N} w_i \right);$

(iv) $s^2 = \overline{w^2} + \bar{w}^2 - 2\bar{w}^2 = \overline{w^2} - \bar{w}^2$

7.10 Skew $= -0.898$

7.11 (i) Organic $C = 0.011 \times$ calcium carbonate $+ 0.606$; (ii) See Fig. b14; (iii) Regression is a poor fit to the data

7.12 Subsidence at Tona $= 1.16 \times$ subsidence at Puig d'Olena $- 25.0$

7.13

	Mount Monger	Emu
Mean	63.87	65.61
Number of values	9	10
Standard deviation	2.60	3.22
Standard error	0.87	1.02
95% confidence	2.00	2.31

Hence, mean SiO_2 content is $63.9 \pm 2.0\%$ at Mount Monger and $65.6 \pm 2.3\%$ at Emu. The large overlap implies that there is no significant difference in mean SiO_2 content in the two locations

Chapter 8

8.1 Gradient is about -30% per km

8.2 Gradients are 4 and 2000 at $x = 2$ and $x = 1000$, respectively

8.3 If $\qquad y = x^3$ $\hfill (1)$

then $\quad y + \Delta y = (x + \Delta x)^3$

$\qquad\qquad = x^3 + 3x^2\Delta x + 3x\Delta x^2 + \Delta x^3$ $\hfill (2)$

subtracting (1) from (2) gives $\Delta y = 3x^2\Delta x + 3x\Delta x^2 + \Delta x^3$

dividing by Δx gives $\Delta y/\Delta x = 3x^2 + 3x\Delta x + \Delta x^2$

as Δx tends to zero this becomes $dy/dx = 3x^2$

8.4 If $y = x^3$ then this is of the form $y = x^n$ with $n = 3$. Hence, gradient is $dy/dx = nx^{n-1} = 3x^2$

8.5 (i) $dy/dx = 20x^{19}$; (ii) $dw/dz = e^z$; (iii) $dw/dx = zx^{z-1}$

8.6 (i) $dy/dx = 10x^9 + 1/\cos^2(x)$; (ii) $dy/dx = 2x + 3x^2 + 4x^3$; (iii) $dw/dz = 75e^z$; (iv) $d\alpha/d\theta = 10\cos(\theta) - 13\sin(\theta)$

8.7 $dT/dz = 4az^3 + 3bz^2 + 2cz + d$

with $a = -1.12 \times 10^{-12}$, $b = 2.85 \times 10^{-8}$, $c = -0.000\,310$, $d = 1.64$, $z = 1000$ gives $dT/dz = 1.1°/km$

8.8 (i) $d\alpha/dx = x^2.e^x + 2x.e^x$; (ii) $dy/dw = 6w.\sin(w) + 3w^2\cos(w)$; (iii) $dz/dx = \cos(x) - x\sin(x) + 3x^2.\tan(x) + x^3/\cos^2(x)$; (iv) $dB/d\sigma = 12\,\sigma^3.\ln(\sigma) + 3\sigma^3 + 34\sigma$

8.9 $u = x^4$ gives $du/dx = 4x^3$

$v = \sin(x)$ gives $dv/dx = \cos(x)$

hence, $\dfrac{dy}{dx} = \dfrac{v(du/dx) - u(dv/dx)}{v^2} = (\sin(x).4x^3 - x^4.\cos(x))/\sin^2(x)$

8.10 $x = \ln(y^2)$ (1)

let $u = y^2$ (2)

then $x = \ln(u)$ (3)

and $du/dy = 2y$ (4)

equation (3) gives $dx/du = 1/u$ (5)

chain rule is $dx/dy = dx/du \, . \, du/dy = (1/u).(2y)$

$\qquad\qquad\qquad\qquad = (1/y^2).(2y) = 2/y$

8.11 The measurements differ by less than their uncertainties after altitude correction. Hence, differences are not significant

8.12 $dy/dx = 6x + 2$

$\qquad d^2y/dx^2 = 6$

$\qquad d^3y/dx^3 = 0$

$\qquad d^4y/dx^4 = 0$

8.13 (i) $S = S_{max}(1 - e^{-0/\tau}) = S_{max}(1 - 1) = 0$;   (ii) $S = S_{max}(1 - e^{-\infty/\tau}) = S_{max}(1 - 0) = S_{max}$; (iii) $S = S_{max} - S_{max}e^{-t/\tau}$, therefore $dS/dt = 0 + (S_{max}/\tau)e^{-t/\tau}$. At $t = 0$ this gives $dS/dt = S_{max}/\tau$; (iv) Result from (iii) gives $dS/dt = 0$ at $t = \infty$; (v) Result from (iii) gives $dS/dt = 3/50 = 0.06$ km/My or 60 m/My

8.14 (i) At $x = X$ salinity equation gives $3s_0 = s_0 \, \alpha X/(\alpha X - X) = s_0 \, \alpha/(\alpha - 1)$ Hence $3 = \alpha/(\alpha - 1)$ which can be rearranged to give $\alpha = 1.5$; (ii) If $s_0 = 30$ ppm and $x = X/2$, salinity equation becomes $s = 45X/(1.5X - X/2) = 45/(1.5 - 0.5) = 45$ ppm; (iii) Use quotient rule and then simplify; (iv) If $X = 10\,000$ m then at $x = X/2$, $ds/dx = s/(\alpha X - x) = 45/(1.5 \times 10\,000 - 5000) = 0.0045$ ppm/m. In other words, salinity changes by 4.5 ppm over 1 km. Hence, 1 km seaward of centre salinity is $45 - 4.5 = 40.5$ ppm. Similarly, 1 km shoreward of centre salinity is $45 + 4.5 = 49.5$ ppm

8.15 (i) $du/d\theta = -2 \, \cos\theta/\sin^3\theta$ and $dv/d\theta = 2 \, \sin\theta/\cos^3\theta$; (ii) $d^2w/d\theta^2 = \alpha et[(2 \, \sin\theta/\cos^3\theta) + (2 \, \cos\theta/\sin^3\theta)] = 2\alpha et(\sin^4\theta + \cos^4\theta)/\cos^3\theta.\sin^3\theta$; (iii) If $\cos\theta = \sin\theta = 1/\sqrt{2}$ then $\cos^4\theta = \sin^4\theta = 1/4$, $\cos\theta.\sin\theta = 1/2$ and therefore $\cos^3\theta.\sin^3\theta = 1/8$. Substitute these into the above expression to get a result.

8.16 (i) $dt/dx = (-t_0/X) \exp(-x/X)$;

$\qquad$ (ii) $dt/dx = (-10/5) \exp(-3/5) = -1.098$ m/km

8.17 (i) If $a = k.z$ then $da/dz = k$; (ii) $\Delta a = (da/dz).\Delta a = k\Delta a$;

$\qquad$ (iii) $\Delta a = 3000 \times 0.1 = 300$ m

8.18 $h = x.\tan(\alpha) = 100.\tan(20°) = 36.4$ m

$dh/d\alpha = x/\cos^2(\alpha)$ Hence height error, $\Delta h = (dh/da).\Delta\alpha = \Delta\alpha \, x/\cos^2(\alpha)$

Angle error, $\Delta\alpha = 2° \times \pi/180° = 0.035$ radians.

Therefore, $\Delta h = 0.035 \times 100/\cos^2(20°) = 3.96$ m

Hence, cliff height is 36.4 ± 4.0 m

Chapter 9

9.1 $\int 10e^{2\alpha}d\alpha = 5e^{2\alpha} + k$

9.2 (i) $\int \cos(x)\, dx = \sin(x) + k$; (ii) $\int \xi^{10}\, d\xi = (\xi^{11}/11) + k$

9.3 $\int_0^X x^2 dx = \left[\dfrac{x^3}{3}\right]_0^X = \dfrac{X^3}{3} - \dfrac{0}{3} = \dfrac{X^3}{3}$

9.4 $\int_0^X x^n dx = \dfrac{X^{n+1}}{n+1}$

For example $n = 3$, $X = 10$ gives $Area = 2500$

9.5 6.873

9.6 $10\ln(x) + \sin(x) + k$

9.7 Thickness $= a_0 + a_1 x + a_2 x^2 + a_3 x^3 + a_4 x^4$ with $a_0 = -0.0778$, $a_1 = 0.0\,142$, $a_2 = -2.17 \times 10^{-5}$, $a_3 = 1.30 \times 10^{-8}$, $a_4 = -2.86 \times 10^{-12}$
Integration gives $Area = a_0 X + (a_1 X^2/2) + (a_2 X^3/3) + (a_3 X^4/4) + (a_4 X^5/5) = 4180$ m^2

9.8 (i) See Fig. b15; (ii) $Q.\Delta z$; (iii) $\sum\limits_{i=1}^n Q_i \Delta z = \sum\limits_{i=1}^n \dfrac{y_i}{20}\Delta z$; (iv) $\dfrac{1}{20}\int_0^{30} y dz$;
(v) 22.5 kW km^{-2}; (vi) 4444 km^2

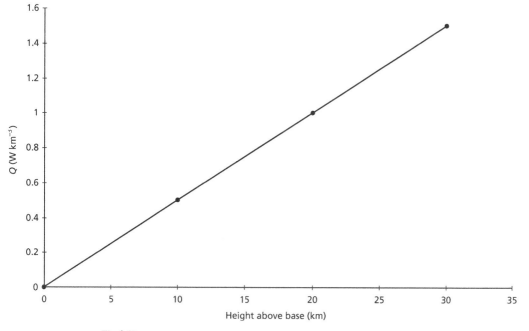

Fig. b15

9.9 (i) $\pi r^2 \, \Delta z$; (ii) $\displaystyle\sum_{i=1}^{n} \pi r_i^2$; (iii) $\displaystyle\int_{-r_p}^{r_p} \pi r^2 dr = \pi \int_{-r_p}^{r_p} r_e^2[1 - (z^2/r_p^2)]dz$

$$= \pi r_e^2\left[2r_p - \frac{2}{3}r_p\right] = \frac{4}{3}\pi r_e^2 r_p;$$ (iv) 1.083×10^{12} km^3; (v) 1.081×10^{12} km^3

9.10 (i) $t_i \Delta x$; (ii) $\displaystyle\sum_{i=1}^{N} t_i \Delta x$; (iii) Xt_0; (iv) 500 years

Index

Note: page numbers in *italic* type refer to figures; those in **bold** type refer to tables

addition
  in scientific notation 11–12
  vectors 83–5
AFM diagrams 95, 97
allometric growth laws *see* exponential
    functions
altitude, and gravity 41, 151–2
angles
  Cartesian coordinates 76–81
  determination 74–6
  measures 69
  trigonometric functions 76–81
animal mass, and foot size 91–5
answers, checking 62–5
approximations
  and assumptions 4
  mathematics as 2–5, 25
  and quality assurance 62–3
arguments, logarithms 56
assumptions, and approximations 4
averages 176
  mean 114, 176
  median 114
  of sample 112

bar charts, and histograms 118
barred basin model, of evaporite formation
    157–8
best fit, straight lines 124–7, 177
bias, statistical 116
bottomset beds, thickness 66, 158–9,
    175
brackets
  and factorization 47–50
  manipulation 47–50
browser software 14
burial, rates 136
burial depth *see* depth

calcium carbonate, content analyses 133
calculators, and scientific notation 13
calculus
  applications 150–2
  discovery 136

see also differential calculus; integral
    calculus
carbonate sediments, accumulation 41
Cartesian coordinates
  angles 76–81
  origin 76
chain rule, differentiation 148–50, 177
classes, statistical 118
classification schemes, end members 98
combinations, and probability 111
common factors 49
common logarithms 38
compaction, rates 137
complex expressions
  differentiation 145–7
  integration 166–8
component vectors 85–8
computers, and mathematics 14
constant of integration 163, 164, 167
cosine function
  and Cartesian coordinates 78
  definition 71
  differentiation 146, 152–3
  integration 163, 164–5, 166, 167
  inverse 73, 80
cosine rule 75, 176
crustal heat, determination 171
crustal thickness, and mountain height
    16
cubes, stereographic projections 107
curves
  areas under 160–2
  tangents 138–9
cylindrical folds 108

dating, radioactive 57–8, 65
definite integration 165–6
degrees, as angular measures 69
degrees of freedom 130
  concept of 131
deltas
  bottomset beds 66, 158–9, 175
  classification 98
density
  and depth 91, 92
  Earth 45–7, 63–4, 172–4
  rocks 53–4